HEINEMANN MODULAR MATHEMATICS
for
EDEXCEL AS AND A-LEVEL
Decision Maths 2

John Hebborn

2, 30, 88

transportation, allocation, game theory

Heinemann

Edexcel
Success through qualifications

Heinemann Educational Publishers,
a division of Heinemann Publishers (Oxford) Ltd,
Halley Court, Jordan Hill, Oxford, OX2 8EJ

OXFORD   JOHANNESBURG   BLANTYRE   MELBOURNE
AUCKLAND   GABORONE   PORTSMOUTH NH (USA)   CHICAGO

First published 2001

05   04   03   02   01
10   9   8   7   6   5   4   3   2   1

ISBN 0 435 51081 9

Cover design by Gecko Limited

Original design by Geoffrey Wadsley: additional design work by Jim Turner

Typeset and illustrated by Tech-Set Limited, Gateshead, Tyne and Wear

Printed in Great Britain by Scotprint

Acknowledgements:

The publisher's and author's thanks are due to Edexcel for permission to
reproduce questions from past examination papers. These are marked with an [E].
The answers have been provided by the author and are not the responsibility of
the examining board.

About this book

This book is designed to provide you with the best preparation possible for your Edexcel D2 exam. The series authors are senior examiners and exam moderators themselves and have a good understanding of Edexcel's requirements.

Finding your way around

To help to find your way around when you are studying and revising, use the:

- **edge marks** (shown on the front page) – these help you to get to the right chapter quickly;
- **contents list** – this lists the headings that identify key syllabus ideas covered in the book so you can turn straight to them;
- **index** – if you need to find a topic the **bold** number shows where to find the main entry on that topic.

Remembering key ideas

We have provided clear explanations of the key ideas and techniques you need throughout the book. Key ideas you need to remember are listed in a **summary of key points** at the end of each chapter and marked like this in the chapters:

- $I_{ij} = C_{ij} - R_i - K_j$

Exercises and exam questions

In this book, questions are carefully graded so they increase in difficulty and gradually bring you up to exam standard.

- **exam-style practice papers** on pages 120 and 123 are designed to help you prepare for the exam itself;
- **answers** are included at the end of the book – use them to check your work.

Contents

3 The travelling salesman problem

4 Game theory

5 Dynamic programming

Transportation problems

In D1 you studied linear programming problems and their solutions using the simplex algorithm. In this chapter and the next one the solutions of two special types of linear programming problems are considered by methods other than the simplex algorithm. These are **transportation problems** and **assignment problems**. Both of these problem types could be solved using the simplex algorithm, but the process would result in very large simplex tableaux and numerous simplex iterations.

Because of the special characteristics of each problem, however, alternative solution methods requiring significantly less mathematical manipulation than the simplex method have been developed.

1.1 The transportation problem

The transportation problem deals with the distribution of goods from several points of supply, such as factories, often known as **sources**, to a number of points of demand, such as warehouses, often known as **destinations**.

Each source is able to supply a fixed number of units of the product, usually called the **capacity** or **availability**, and each destination has a fixed demand, usually known as the **requirement**. The objective is to schedule shipments from sources to destinations so that the total transport cost is a minimum.

Example 1

A concrete company transports concrete from three plants, 1, 2 and 3, to three construction sites, A, B and C.

The plants are able to supply the following numbers of tons per week:

Plant	Supply (capacity)
1	300
2	300
3	100

The requirements of the sites, in numbers of tons per week, are:

Construction site	Demand (requirement)
A	200
B	200
C	300

The cost of transporting 1 ton of concrete from each plant to each site is shown in the figure below in pounds per ton.

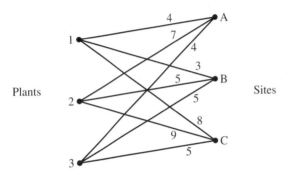

For computational purposes it is convenient to put all the above information into a table, as in the simplex method. In this table each **row** represents a **source** and each **column** represents a **destination**.

Sites

From \ To	A	B	C	Supply (availability)
1	4	3	8	300
2	7	5	9	300
3	4	5	5	100
Demand (requirement)	200	200	300	

Plants

1.2 Formulation as a linear programming problem

Before proceeding with the solution of the transportation problem, using the method developed especially for it, we will show that it can be formulated as a linear programming problem.

The decision variables x_{ij} are the numbers of tons transported from plant i (where $i = 1, 2, 3$) to each site j (where $j = A, B, C$).

The objective function represents the **total transportation cost** £Z. Each term in the objective function Z represents the cost of tonnage transported on one route. For example, for the route $2 \rightarrow C$ the term is $9x_{2C}$, that is:

(cost per ton = 9) × (number of tons transported = x_{2C})

Hence the objective function is:

$$Z = 4x_{1A} + 3x_{1B} + 8x_{1C}$$
$$+7x_{2A} + 5x_{2B} + 9x_{2C}$$
$$+4x_{3A} + 5x_{3B} + 5x_{3C} \tag{1}$$

Notice that in this problem the total supply is $300 + 300 + 100 = 700$ and the total demand is $200 + 200 + 300 = 700$.

So (total supply) = (total demand).

This is called a **balanced problem**. Balanced problems are considered throughout this chapter unless otherwise stated.

In a balanced problem all the products that can be supplied are used to meet the demand. There are no slacks and so all constraints are **equalities** rather than **inequalities**.

The constraints for the problem are obtained by considering the rows and columns of the above table.

Plants: The total number of tons transported from plant 1 to sites A, B and C must be equal to the supply, so:

$$x_{1A} + x_{1B} + x_{1C} = 300 \tag{2a}$$

Similarly, for plants 2 and 3:

$$x_{2A} + x_{2B} + x_{2C} = 300 \tag{2b}$$
$$x_{3A} + x_{3B} + x_{3C} = 100 \tag{2c}$$

Sites: The total number of tons received by site A from plants 1, 2 and 3 must be equal to the demand, so:

$$x_{1A} + x_{2A} + x_{3A} = 200 \tag{3a}$$

Similarly, for sites B and C:

$$x_{1B} + x_{2B} + x_{3B} = 200 \tag{3b}$$
$$x_{1C} + x_{2C} + x_{3C} = 300 \tag{3c}$$

All decision variables must be non-negative, so:

$$x_{ij} \geqslant 0 \tag{4}$$

for $i = 1, 2, 3$ and $j = A, B, C$.

The linear programming problem is then:

Minimise Z subject to the constraints given by equations (2), (3) and (4).

Exercise 1A

1 A steel company has three mills, M_1, M_2 and M_3, which can produce 40, 10 and 20 kilotonnes of steel each year. Three customers, C_1, C_2 and C_3, have requirements of 12, 18 and 40 kilotonnes respectively in the same period. The cost, in units of £1000, of transporting a kilotonne of steel from each mill to each customer is shown in the figure below.

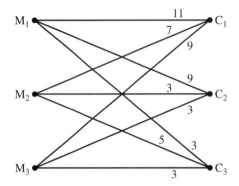

(a) Summarise all the above information in the form of a table.

(b) Formulate the problem of carrying out the transportation at minimum cost as a linear programming problem.

2 A transportation problem involves the following costs, supply and demand:

	W_1	W_2	W_3	Supply
F_1	7	8	6	4
F_2	9	2	4	3
F_3	5	6	3	8
Demand	2	9	4	

Formulate this problem as a linear programming problem.

3 Downside Mills produces carpet at plants in Abbeyville (A) and Bridgeway (B). The carpet is shipped to two outlets in Courtney (C) and Dove Valley (D). The cost, in £ per ton, of shipping carpet from each of the two plants to each of the two outlets is as follows:

From \ To	C	D
A	20	32
B	35	15

The plant at Abbeyville can supply 250 tons of carpet per week and the plant at Bridgeway can supply 300 tons of carpet per week. The Courtney outlet has a demand for 320 tons per week and the outlet at Dove Valley has a demand for 230 tons per week. The company wants to know the number of tons of carpet to ship from each plant to each outlet in order to minimise the total shipping cost. Formulate this transportation problem as a linear programming problem.

1.3 Solution of the transportation problem

Setting up a transportation tableau

The purpose of the transportation tableau is to summarise conveniently and concisely **all** relevant data and keep track of algorithm computations. In this respect it serves the same role as the simplex tableau did for linear programming problems.

The transportation tableau for Example 1 is:

From \ To	A	B	C	Supply (availability)
1	4	3	8	300
2	7	5	9	300
3	4	5	· 5	100
Demand (requirement)	200	200	300	700 ← Total demand and supply

Each cell in the tableau represents the amount transported from one source to one destination. The amount placed in each cell is therefore the value of a decision variable for the cell. For example, the cell at the intersection of row 2 and column C represents the decision variable x_{2C}. The smaller box within each cell contains the unit transportation cost for that route. The final column and the final row are the capacities of the plants and the requirements of the sites respectively. The total demand and total supply has been placed in the bottom right-hand corner. It is suggested that you should always do this to check that you have a balanced problem. Later we will see how to deal with unbalanced problems where total destination requirements are greater than or less than total source capacities.

The transportation algorithm

The transportation algorithm consists of three stages.

Stage 1 Find a transportation pattern that uses all the products available and satisfies all the requirements. This is called **developing an initial solution**.

Stage 2 **Test the solution for optimality**. If the solution is optimal **stop**. If the solution is not optimal move to stage 3.

Stage 3 Use the **stepping-stone method to obtain an improved solution** and then return to stage 2.

The algorithm stops when no further improvement is possible.

Finding an initial solution

An initial feasible solution can be found by several alternative methods. We will give just one, known as the **north-west corner method**.

You will recall that prior to applying the simplex algorithm an initial solution had to be established in the initial simplex tableau. The initial solution in the transportation problem is not the origin. The north-west corner method requires that we start in the upper left-hand cell or **north-west corner** of the table and allocate units to shipping routes as follows:

(i)   Exhaust the supply at each row before moving down to the next row.

(ii) Exhaust the requirements of each column before moving to the right to the next column.

Let us implement this rule for Example 1.

From \ To	A	B	C	Supply
1	200	100		300
2		100	200	300
3			100	100
Demand	200	200	300	

The figures in the above table were obtained in the following way:

(i)   Site A requires 200. All of these can be supplied by plant 1. Plant 1 still has 100 left. These are sent to site B.

(ii) We now move to line 2. Site A is satisfied. Site B still requires 100. These can be sent from plant 2. Plant 2 still has 200 left. These are sent to site C.

(iii) Site C still requires 100. These can be sent from plant 3.

This solution is feasible since supply and demand constraints are all satisfied.

This solution corresponds to $x_{1A} = 200$, $x_{1B} = 100$, $x_{2B} = 100$, $x_{2C} = 200$, $x_{3C} = 100$ and all other variables zero.

The total cost of this solution is:

$$[4 \times (200)] + [3 \times (100)] + [5 \times (100)] + [9 \times (200)] + [5 \times (100)]$$
$$= 800 + 300 + 500 + 1800 + 500$$
$$= \text{£}3900$$

Example 2

Use the north-west corner rule to obtain an initial solution for the transportation tableau below.

Site

From \ To	A	B	C	Supply
1	**40** 10	12	9	40
2	**30** 4	**20** 5	7	50
3	11	**30** 8	**30** 6	60
Demand	70	50	30	150

Plant (labels rows 1, 2, 3)

Note: the answers are shown by the numbers in bold type

(i)   Not all of site A's demand (70) can be met by plant 1. Site A can receive 40 from plant 1 and plant 2 can provide the remaining 30. (It is a good idea to fill in the cells as you go along.)

(ii) Plant 2 still has 20 left and these can be sent to site B.

(iii) Site B still requires 30 and these can be provided by plant 3. Plant 3 still has 30 left and these can be sent to site C to exactly meet its requirement.

The total cost of this solution is:

$$(40 \times 10) + (30 \times 4) + (20 \times 5) + (30 \times 8) + (30 \times 6)$$
$$= 400 + 120 + 100 + 240 + 180$$
$$= 1040$$

Notice that at each step a row or column constraint is satisfied except at the last stage when both a row and a column constraint are satisfied.

An initial solution obtained using the north-west corner rule is often called the **north-west corner solution**. As illustrated in the example below, we can obtain initial solutions by starting at other corners of the tableau.

Example 3

For the transportation tableau below write down:
(a) the north-west corner solution
(b) the north-east corner solution
(c) the south-west corner solution
(d) the south-east corner solution.

Assuming one of these gives an optimal solution, obtain the optimal solution.

	A	B	
1	40	65	250
2	70	30	400
	300	350	

(a) The north-west corner solution obtained using the rule given earlier is:

	A	B	
1	250		250
2	50	350	400
	300	350	

(b) The north-east corner solution is obtained by starting in the top right-hand corner and is:

	A	B	
1		250	250
2	300	100	400
	300	350	

(c) The south-west corner solution is obtained by starting in the bottom left-hand corner and is:

	A	B	
1		250	250
2	300	100	400
	300	350	

This is the same as (b).

(d) The south-east corner solution is obtained by starting in the bottom right-hand corner and is:

	A	B	
1	250		250
2	50	350	400
	300	350	

This is the same as (a).

The cost of (a) is:

$$(250 \times 40) + (50 \times 70) + (350 \times 30) = 24\,000$$

The cost of (b) is:

$$(250 \times 65) + (300 \times 70) + (100 \times 30) = 40\,250$$

So the optimal solution is:

$$250 \text{ from } 1 \rightarrow A$$
$$50 \text{ from } 2 \rightarrow A$$
$$350 \text{ from } 2 \rightarrow B$$

The minimum cost is 24 000.

Exercise 1B

1 Use the north-west corner rule to obtain an initial solution for the following transportation tableaux:

(a)

				Supply
				100
				40
				50
Demand	60	60	70	

(b)

			Supply
			60
			40
			140
Demand	120	80	40

(c)

			Supply
			60
			100
			80
Demand	60	70	110

(d)

			Supply
			90
			40
			50
Demand	60	70	50

2 For the transportation tableau below write down:
(a) the north-west corner solution
(b) the north-east corner solution
(c) the south-west corner solution
(d) the south-east corner solution.

Given that one of these gives an optimal solution, obtain the optimal solution.

	A	**B**	Supply
1	8	6	100
2	10	12	150
Demand	90	160	

1.4 Testing the solution for optimality

An important rule

The method we shall now describe for testing a solution for optimality can only be applied if one essential condition is satisfied.

This condition is:

- **The number of occupied cells (routes used) must be equal to one less than the sum of the number of rows and the number of columns.**

In both Example 1 and Example 2 we have three rows and three columns. Since the number of occupied cells is 5 in each case and this is equal to $(3 + 3 - 1)$, the condition is satisfied.

In the general case, when we have m sources and n destinations the number of occupied cells must be $(m + n - 1)$.

When the number of occupied cells is less than this the solution is said to be **degenerate**. Later in this chapter we will consider what to do when we have degeneracy.

Calculation of improvement indices

To test a solution for optimality we need to calculate an **improvement index** for each unused cell. As a first step in this process we must compute:
(i) a value for each row, denoted by R_i
(ii) a value for each column, denoted by K_j.

If C_{ij} is the unit cost in the cell in the ith row and jth column of the transportation tableau, then we can obtain the above values by using:

- $$R_i + K_j = C_{ij}$$

for the **occupied (used) cells**. The R_i and K_j are called **shadow costs**.

The north-west corner solution for Example 1 is repeated below.

To / From	A	B	C	Supply
1	200 ⌐4	100 ⌐3	⌐8	300
2	⌐7	100 ⌐5	200 ⌐9	300
3	⌐4	⌐5	100 ⌐5	100
Demand	200	200	300	700

So for this solution we have:

$$\text{cell } (1, 1): \quad R_1 + K_1 = 4 \tag{1}$$
$$\text{cell } (1, 2): \quad R_1 + K_2 = 3 \tag{2}$$
$$\text{cell } (2, 2): \quad R_2 + K_2 = 5 \tag{3}$$
$$\text{cell } (2, 3): \quad R_2 + K_3 = 9 \tag{4}$$
$$\text{cell } (3, 3): \quad R_3 + K_3 = 5 \tag{5}$$

Here we have five equations in six unknowns: R_1, R_2, R_3, K_1, K_2, and K_3. **We may therefore choose one of the unknowns and solve for the others**. It is usual to choose $R_1 = 0$, but you do not have to, you may choose any one of the unknowns.
Then from:

(1) $$0 + K_1 = 4 \Rightarrow K_1 = 4$$
(2) $$0 + K_2 = 3 \Rightarrow K_2 = 3$$
(3) $$R_2 + K_2 = R_2 + 3 = 5 \Rightarrow R_2 = 2$$
(4) $$R_2 + K_3 = 2 + K_3 = 9 \Rightarrow K_3 = 7$$
(5) $$R_3 + K_3 = R_3 + 7 = 5 \Rightarrow R_3 = -2$$

It is always possible to solve the resulting set of equations in this step-by-step way but the calculations may not proceed from equation (1) $\rightarrow$ equation (2) $\rightarrow$ equation (3) $\rightarrow$ equation (4) $\rightarrow$ equation (5) as above. Notice that it is possible that the R_i and K_j values may be positive, negative or zero. After solving for the R and K values a few times you may become so proficient that you will be able to do the calculations in your head. The following table summarises the results and may be written down straight away if you do the calculations in your head. This table is also useful in calculating improvement indices.

	$K_1 = 4$	$K_2 = 3$	$K_3 = 7$
$R_1 = 0$	X 4	X 3	8
$R_2 = 2$	7	X 5	X 9
$R_3 = -2$	4	5	X 5

Notice that we have indicated the used cells by X and only included the costs. Do not get confused: it is the **costs** that are involved in the calculations of the R and K values and only the costs. Some students do get confused on this matter.

Having calculated the R and K values we can now calculate, **for each unused cell**, an improvement index I_{ij} using the formula:

$$\blacksquare \qquad I_{ij} = C_{ij} - R_i - K_j$$

By definition of the R_i and K_j values the improvement index for an occupied (used) cell is zero.

For the north-west corner solution for Example 1 we have, using the table above:

cell (1, 3): $I_{13} = C_{13} - R_1 - K_3 = 8 - 0 - 7 = 1$
cell (2, 1): $I_{21} = C_{21} - R_2 - K_1 = 7 - 2 - 4 = 1$
cell (3, 1): $I_{31} = C_{31} - R_3 - K_1 = 4 - (-2) - 4 = 2$
cell (3, 2): $I_{32} = C_{32} - R_3 - K_2 = 5 - (-2) - 3 = 4$

A warning: Do be careful when you are subtracting negative numbers. For example, $2 - (-2) = +4$.

Again, with a little practice these calculations can be done in your head and the results written into a table such as that below.

Improvement indices

	$K_1 = 4$	$K_2 = 3$	$K_3 = 7$
$R_1 = 0$	⓪ 4	⓪ 3	1 8
$R_2 = 2$	1 7	⓪ 5	⓪ 9
$R_3 = -2$	2 4	4 5	⓪ 5

Notice that the occupied cells have been indicated with ringed zeros.

■ **If all the improvement indices are greater than or equal to zero, an optimal solution has been reached.**

If there are any negative improvement indices then it is possible to **improve** the current solution and **decrease** the total shipping costs. Each negative index computed represents the amount by which the total transportation costs could be decreased if 1 unit were shipped on that route.

In the light of the above, the north-west corner solution to Example 1 is **optimal**. The solution to the transportation problem posed in Example 1 is:

send   200 tons from plant 1 to site A
                          100 tons from plant 1 to site B
                          100 tons from plant 2 to site B
                          200 tons from plant 2 to site C
                          100 tons from plant 3 to site C

The cost of the transportation pattern, which is the minimum total cost, is £3900, as shown earlier.

Example 4
Due to roadworks the cost of shipping 1 ton from plant 1 to site A in Example 1 increases to £6. All the remaining data remain the same. Obtain the improvement indices for the north-west corner solution now.

The north-west corner solution is:

To From	A	B	C	Supply
1	200 ⌐6	100 ⌐3	⌐8	300
2	⌐7	100 ⌐5	200 ⌐9	300
3	⌐4	⌐5	100 ⌐5	100
Demand	200	200	300	700

The equations to be solved for the R and K values are now:

$$R_1 + K_1 = 6$$
$$R_1 + K_2 = 3$$
$$R_2 + K_2 = 5$$
$$R_2 + K_3 = 9$$
$$R_3 + K_3 = 5$$

Taking $R_1 = 0$, we obtain $R_2 = 2$, $R_3 = -2$, $K_1 = 6$, $K_2 = 3$ and $K_3 = 7$.

The improvement indices for the non-occupied cells are then:

$$I_{13} = 8 - 0 - 7 = 1$$
$$I_{21} = 7 - 2 - 6 = -1$$
$$I_{31} = 4 - (-2) - 6 = 0$$
$$I_{32} = 5 - (-2) - 3 = 4$$

As there is now a negative improvement index the solution is no longer optimal and can be improved.

Exercise 1C

1 In Example 2 we considered the transportation problem given by the following table of supply, demand and unit costs:

To From	A	B	C	Supply
1	10	12	9	40
2	4	5	7	50
3	11	8	6	60
Demand	70	50	30	

We showed in Example 2 that the north-west corner rule gave the following initial solution:

To From	A	B	C	Supply
1	40 $\boxed{10}$	$\boxed{12}$	$\boxed{9}$	40
2	30 $\boxed{4}$	20 $\boxed{5}$	$\boxed{7}$	50
3	$\boxed{11}$	30 $\boxed{8}$	30 $\boxed{6}$	60
Demand	70	50	30	150

By calculating improvement indices for the unoccupied cells, show that this solution is optimal. Give the transportation pattern and its cost.

2 A company has three factories, F_1, F_2 and F_3, and three warehouses, W_1, W_2 and W_3. The table below shows the costs C_{ij} of sending one unit of product from factory F_i to warehouse W_j. Also shown are the availabilities at each factory and the requirements of each warehouse.

To From	W_1	W_2	W_3	Availability
F_1	8	6	7	4
F_2	2	4	9	3
F_3	6	3	5	8
Requirement	9	4	2	

(a) Use the north-west corner rule to write down a possible pattern of distribution and find its cost.
(b) Calculate the improvement indices for the unused cells and hence show that this distribution pattern is optimal.

3 A company has three warehouses, W_1, W_2 and W_3, which are supplied by three suppliers, S_1, S_2 and S_3. The table below shows the cost C_{ij} of sending one case of goods from supplier S_i to warehouse W_j, in appropriate units. Also shown in the table are the number of cases available at each supplier and the

number of cases required at each warehouse. The total number of cases available is equal to the total number of cases required.

	W_1	W_2	W_3	Availability
S_1	10	4	11	14
S_2	12	5	8	10
S_3	9	6	7	6
Requirement	8	10	12	

(a) Use the north-west corner rule to obtain a possible pattern of distribution.

(b) Calculate the improvement indices for the unused cells and hence show that this distribution pattern is not optimal.

Obtaining an improved solution (the stepping-stone method)

Now that we have decided that the north-west corner solution in Example 4 can be improved by using route (2, 1), we must decide which route is to be removed from the present solution. This procedure is similar to that used in the simplex algorithm, where we first choose an entering variable and then choose a departing variable.

In order to determine the cell to be used we need to draw a **closed path** or **loop** in the transportation tableau.

■ **A closed path or loop is a sequence of cells in the transportation tableau such that:**
   (i) each pair of consecutive cells lies in either the same row or the same column
   (ii) no three consecutive cells lie in the same row or column
   (iii) the first and last cells of a sequence lie in the same row or column
   (iv) no cell appears more than once in the sequence.

The loops we are interested in have the following property:
the first cell is unused and all the other cells are used.
For Example 4 above the used cells are shown by X in the table below:

X	X	
	X	X
		X

The possible loops with the above property are:

X⌐ ¬X		
⌐¬X	X	
	X	

(i) starting at (2, 1)

X	X⌐ ¬	
	X⌐¬X	
	X	

(ii) starting at (1, 3)

X	X	
	X⌐ ¬X	
	⌐¬X	

(iii) starting at (3, 2)

X ⌐ ¬X		
	X⌐ ¬X	
	⌐———¬X	

(iv) starting at (3, 1)

It should be noted that in developing the loop it is possible to skip over both unused and used cells.

Since we wish to use cell (2, 1), we now concentrate our attention on diagram (i) above. To ensure that the constraints remain satisfied, we add θ to the number in (2, 1) and $(-\theta)$ and $(+\theta)$ to the numbers in the other cells, alternately, as we move round the loop. We then have:

	A	**B**	**C**	
1	$200 - \theta$	$100 + \theta$		300
2	$+\theta$	$100 - \theta$	200	300
3			100	100
	200	200	300	

You may check each row and column in turn to see that the constraints are satisfied.

To satisfy the non-negativity condition and improve the solution as much as possible **we make θ as large as possible without making any entry negative**. So choose $\theta = \min(100, 200) = 100$.

Taking $\theta = 100$ gives us the new improved solution:

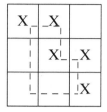

Notice that route (2, 2) is not now used.

We will now test this solution for optimality.

The equations to be solved for the R and K values are:

$R_1 + K_1 = 6$
$R_1 + K_2 = 3$
$R_2 + K_1 = 7$
$R_2 + K_3 = 9$
$R_3 + K_3 = 5$

	$K_1 = 6$	$K_2 = 3$	$K_3 = 8$
$R_1 = 0$	X ⌐6	X ⌐3	⌐8
$R_2 = 1$	X ⌐7	⌐5	X ⌐9
$R_3 = -3$	⌐4	⌐5	X ⌐5

Taking $R_1 = 0$, we obtain $R_2 = 1$, $R_3 = -3$, $K_1 = 6$, $K_2 = 3$ and $K_3 = 8$ (see table).

The improvement indices for the unused cells are:

$$I_{13} = 8 - 0 - 8 = 0$$
$$I_{22} = 5 - 1 - 3 = 1$$
$$I_{31} = 4 - (-3) - 6 = 1$$
$$I_{32} = 5 - (-3) - 3 = 5$$

All improvement indices are non-negative and so this solution is optimal:

send   100 tons from plant 1 to site A
            200 tons from plant 1 to site B
            100 tons from plant 2 to site A
            200 tons from plant 2 to site C
            100 tons from plant 3 to site C

Take cost of this transportation pattern is:

$$(100 \times 6) + (200 \times 3) + (100 \times 7) + (200 \times 9) + (100 \times 5)$$
$$= 600 + 600 + 700 + 1800 + 500$$
$$= £4200$$

Alternative optimal solutions

You may have noticed that one of the improvement indices calculated above (I_{13}) was zero. This indicates that the route (1, 3) could be used **without changing the total overall cost**:

X	X	
X		X
		X

The routes used in the optimal solution obtained are shown by X in the above diagram. Also shown is the closed path or loop starting at (1, 3). Proceeding as above we have:

$100 - \theta$	200	$+\theta$
$100 + \theta$		$200 - \theta$
		100

In this case $\theta = \min(100, 200) = 100$, so we have the alternative solution:

	6	200	3	100	8
200	7		5	100	9
	4		5	100	5

send
200 tons from plant 1 to site B
100 tons from plant 1 to site C
200 tons from plant 2 to site A
100 tons from plant 2 to site C
100 tons from plant 3 to site C

The total cost of this transportation pattern is:

$$(200 \times 3) + (100 \times 8) + (200 \times 7) + (100 \times 9) + (100 \times 5)$$
$$= 600 + 800 + 1400 + 900 + 500$$
$$= £4200, \text{ as before}$$

Exercise 1D

1 In question 3 of Exercise 1C you showed that the solution obtained by the north-west corner rule was not optimal. Obtain an improved solution using the stepping-stone method and show that this solution is optimal.

2 A lumber company ships pine flooring from its three mills, A_1, A_2 and A_3, to three building suppliers, B_1, B_2 and B_3. The table below shows the demand, availabilities and unit costs of transportation. Starting with the north-west corner solution and using the stepping-stone method, determine the transportation pattern that minimises the total cost.

	B_1	B_2	B_3	Availability
A_1	3	3	2	25
A_2	4	2	3	40
A_3	3	4	3	31
Demand	30	30	36	

1.5 Some special situations

Unbalanced transportation problems

So far in this chapter we have considered only **balanced transportation problems** in which (total supply) = (total demand).

In many real-world situations, however, supply exceeds demand, or vice versa. When we have an **unbalanced transportation problem** we need to convert it into a balanced transportation problem in order to use the solution procedure described earlier in this chapter.

Example 5

Consider the problem given in Example 3 but with source 1 now having 350 products available rather than 250. The transportation tableau for this problem is now obtained by adding a dummy column, that is a dummy destination.

This destination is assigned a dummy requirement equal to the surplus available, in this case $(400 + 350) - (300 + 350) = 100$. The transportation costs in the dummy column are zero, as these surplus products will not, of course, be transported.

The transportation tableau is now:

To / From	A	B	Dummy	Supply
1	40	65	0	350
2	70	30	0	400
Demand	300	350	100	750

We can now apply the transportation algorithm described earlier. The north-west corner solution is:

300 40	50 65	0	350
70	300 30	100 0	400
300	350	100	750

The R and K values are:

	$K_1 = 40$	$K_2 = 65$	$K_3 = 35$
$R_1 = 0$	X 40	X 65	0
$R_2 = -35$	70	X 30	X 0

The improvement indices for the unused cells are:

$$I_{13} = 0 - 0 - 35 = -35$$
$$I_{21} = 70 - (-35) - 40 = 65$$

Since $I_{13} = -35$, that is it is negative, we can improve the solution by using the route (1, 3).

The required loop is:

X	X	
	X	X

Adding and subtracting θ as required gives:

300	$50 - \theta$	θ
	$300 + \theta$	$100 - \theta$

In this case $\theta = \min(50, 100) = 50$ and so we obtain the improved pattern:

300		50
	350	50

The R and K values for this pattern are:

	$K_1 = 40$	$K_2 = 30$	$K_3 = 0$
$R_1 = 0$	X 40	65	X 0
$R_2 = 0$	70	X 30	X 0

and the improvement indices for the unused cells are:

$$I_{12} = 65 - 0 - 30 = 35$$
$$I_{21} = 70 - 0 - 40 = 30$$

Since all improvement indices are positive, this pattern is optimal.

Send   300 tons from 1 to A
                 350 tons from 2 to B

There will be 50 products left at 1 and 50 products left at 2.
The cost of this transportation pattern is:

$$(300 \times 40) + (350 \times 30)$$
$$= 22\,500$$

This is to be compared with the cost of the north-west corner solution, which is:

$$(300 \times 40) + (50 \times 65) + (300 \times 30)$$
$$= 24\,250$$

If the total demand exceeds the total supply, then we add a dummy source with supply = (total demand) − (total supply). A dummy row is added to the tableau and as before the transportation costs in this row will be zero. The transportation algorithm can then be used to solve the problem.

Exercise 1E

1 A manufacturing company has three factories, F_1, F_2 and F_3, and two retail outlets, R_1 and R_2. It wishes to transport its products from its factories to its outlets at minimum total cost. The table below gives details of demand and supply, and also the unit costs of transportation.

From \ To	R_1	R_2	Supply
F_1	2	6	30
F_2	2	4	60
F_3	6	9	20
Demand	60	20	

(a) Write down the north-west corner solution to the corresponding balanced problem.

(b) Show that this solution is optimal.

(c) State the optimal transportation pattern and gives its cost.

2 A transportation problem involves the following costs, supply and demand:

From \ To	D_1	D_2	D_3	Supply
S_1	7	8	10	50
S_2	9	7	8	60
Demand	70	30	40	

(a) Show that the problem is unbalanced and draw a transportation tableau for the corresponding balanced problem.

(b) Write down the north-west corner solution and then use the stepping-stone method to obtain an optimal solution.

(c) Interpret your optimal solution and obtain the minimum cost.

Degeneracy

■ **A feasible solution to a transportation problem is degenerate if the number of used cells is less than the magic number ($m + n - 1$).**

Degeneracy can occur during the determination of the initial feasible solution or it can occur during subsequent iterations when the stepping-stone method is used.

Degeneracy requires a special procedure to deal with the problem. To handle degenerate problems we **create an artificially occupied cell**, that is we place a zero in one of the unused cells and then treat that cell as if it were occupied. The cell must be chosen so that it is possible to calculate all the R and K values. There is usually some flexibility in selecting the unused cell that will receive the zero.

Degeneracy in an initial solution

Consider the following transportation tableau:

To From	A	B	C	Supply
1	5	4	12	200
2	8	6	10	100
3	11	7	11	200
Demand	100	200	200	500

Using the north-west corner method gives the initial solution:

To From	A	B	C	Supply
1	100 5	100 4	12	200
2	8	100 6	10	100
3	11	7	200 11	200
Demand	100	200	200	500

This initial solution is degenerate since the number of used cells is 4, which is less than the magic number $(3 + 3 - 1) = 5$. The degeneracy arises because when 100 is placed in (2, 2) a row **and** a column requirement are satisfied simultaneously. To deal with this problem we place a zero in an unused cell. In this case we may choose either cell (2, 3) or cell (3, 2).

If we choose to place a zero in (2, 3), then the equations to be solved for R and K are:

$R_1 + K_1 = 5$

$R_1 + K_2 = 4$

$R_2 + K_2 = 6$

$R_2 + K_3 = 10$

$R_3 + K_3 = 11$

	$K_1 = 5$	$K_2 = 4$	$K_3 = 8$
$R_1 = 0$	X [5]	X [4]	[12]
$R_2 = 2$	[8]	X [6]	X [10]
$R_3 = 3$	[11]	[7]	X [11]

Taking $R_1 = 0$, we obtain $R_2 = 2$, $R_3 = 3$, $K_1 = 5$, $K_2 = 4$ and $K_3 = 8$.

The improvement indices for the unused cells are:

$$I_{13} = 12 - 0 - 8 = 4$$
$$I_{21} = 8 - 2 - 5 = 1$$
$$I_{31} = 11 - 3 - 5 = 3$$
$$I_{32} = 7 - 3 - 4 = 0$$

Since all these indices are non-negative, this solution is optimal. Since also $I_{32} = 0$ there is an alternative solution in which route (3, 2) is used. This corresponds to making the alternative choice for the cell to receive the zero.

The optimal solution to this problem is:

send
100 units from 1 to A
100 units from 1 to B
100 units from 2 to B
200 units from 3 to C

The total cost of this pattern is:

$$(100 \times 5) + (100 \times 4) + (100 \times 6) + (200 \times 11)$$
$$= 3700$$

Degeneracy during later solution stages

Consider the transportation tableau:

To From	A	B	C	Supply
1	 2	 4	 5	150
2	 3	 8	 6	200
Demand	100	150	100	350

The initial solution obtained by the north-west corner rule is:

To From	A	B	C	Supply
1	100 2	50 4	 5	150
2	 3	100 8	100 6	200
Demand	100	150	100	350

The R and K values are:

	$K_1 = 2$	$K_2 = 4$	$K_3 = 2$
$R_1 = 0$	X 2	X 4	 5
$R_2 = 4$	 3	X 8	X 6

The improvement indices for the unused cells are:

$$I_{13} = 5 - 0 - 2 = 3$$
$$I_{21} = 3 - 4 - 2 = -3$$

There is a negative improvement index I_{21} and so the solution may be improved:

	A	B	C
1	$100 - \theta$ X	$50 + \theta$ X	
2	$+\theta$	X $100 - \theta$	X

The relevant loop is shown above together with the θ adjustment. In this case we choose $\theta = 100$ and obtain the new solution:

To From	A	B	C	Supply
1	⌐2	150 ⌐4	⌐5	150
2	100 ⌐3	⌐8	100 ⌐6	200
Demand	100	150	100	350

The number of used cells is 3 and since this is less than the magic number $(3 + 2 - 1) = 4$ the solution is degenerate. This is because two formerly occupied cells have dropped to zero. **To proceed we have to add a zero to one of the unoccupied cells and then treat it as occupied**. It is usual to choose the cell with the lowest shipping cost, in this case (1, 1).

The R and K values are:

	$K_1 = 2$	$K_2 = 4$	$K_3 = 5$
$R_1 = 0$	X ⌐2	X ⌐4	⌐5
$R_2 = 1$	X ⌐3	⌐8	X ⌐6

The improvement indices of the unused cells are:

$$I_{13} = 5 - 0 - 5 = 0$$
$$I_{22} = 8 - 1 - 4 = 3$$

Again there is an alternative solution obtained by using (1, 3) instead of (1, 1) as the cell in which to place the zero.
The optimal solution is:

send
150 units from 1 to B
100 units from 2 to A
100 units from 2 to C

Total cost of this pattern is:

$$(150 \times 4) + (100 \times 3) + (100 \times 6)$$
$$= 1500$$

Exercise 1F

1 A manufacturing company produces diesel engines in three cities, C_1, C_2 and C_3, and they are purchased by three trucking companies, T_1, T_2 and T_3. The table below shows the number of engines available at C_1, C_2 and C_3 and the number of engines required by T_1, T_2 and T_3. It also shows the transportation cost per engine (in £100s) from sources to destinations. The company wishes to keep the total transportation costs to a minimum.

From \ To	T_1	T_2	T_3	Supply
C_1	3	2	3	25
C_2	4	2	3	35
C_3	3	2	6	20
Demand	30	30	20	

(a) Write down the north-west corner solution and state why it is degenerate.

(b) Use the stepping-stone method to obtain the optimal solution.

(c) Give the transportation pattern and its total cost.

2 A builders' merchant has 13 tons of sand at site X, 11 tons at site Y and 10 tons at site Z. He has orders for 9 tons from customer A, 13 tons from customer B and 12 tons from customer C. The cost per ton (in £10s) of moving the sand between depots and customers is given in the table below.

	A	B	C
X	1	2	4
Y	1	3	4
Z	5	7	5

(a) Write down the north-west corner solution.

(b) Use the stepping-stone method to obtain an optimal solution. State the minimum cost of transportation.

3 A transportation problem involves the following costs, supply and demand:

	D_1	D_2	D_3	Supply
S_1	17	8	14	30
S_2	15	10	20	20
S_3	20	5	10	10
Demand	10	20	30	

(a) Write down the north-west corner solution.

(b) Use the stepping-stone method to obtain an optimal solution and give its cost.

SUMMARY OF KEY POINTS

1 The number of occupied cells (routes used) must be equal to one less than the sum of the number of rows and the number of columns.

2 The **shadow costs** R_i, for the ith row, and K_j, for the jth column, are obtained by solving $R_i + K_j = C_{ij}$ for **occupied cells**, taking $R_1 = 0$ arbitrarily.

2 The **improvement index** I_{ij} for an **unoccupied cell** is defined by

$$I_{ij} = C_{ij} - R_i - K_j.$$

4 If all improvement indices are **greater than or equal to zero** then an **optimal solution** has been reached.

5 A **closed path** or **loop** is a sequence of cells in the transportation tableau such that
(i) each pair of consecutive cells lies in either the same row or the same column
(ii) no three consecutive cells lie in the same row or column
(iii) the first and last cells of a sequence lie in the same row or column
(iv) no cell appears more than once in the sequence.

6 A feasible solution to a transportation problem is **degenerate** if the number of used cells is less than the magic number $(m + n - 1)$, where m is the number of rows and n is the number of columns.

Allocation (assignment) problems

2

2.1 The assignment problem

The assignment problem arises in a variety of situations. Typical assignment problems are:

> assigning jobs to machines
> assigning sales personnel to sales territories
> assigning contracts to bidders
> assigning agents to tasks
> assigning taxis to customers

The distinguishing feature of assignment problems is that **one agent is assigned to one and only one task**. For example, if we have three taxis, A, B and C, and three customers, 1, 2 and 3, who want a taxi then A → 3, B → 1, C → 2 is a possible assignment.
We will, initially, assume that

$$(\text{number of agents}) = (\text{number of tasks})$$

Problems satisfying this condition are called **balanced problems**. We will consider how to deal with unbalanced problems later in the chapter.

In solving the assignment problem we are looking for a set of pairings, for example agents to tasks, which optimises a stated objective such as minimising total cost or distance or time.

To illustrate the assignment problem consider the following example.

Example 1
A manager has three workers, A, B and C, who are to be assigned to three jobs, 1, 2 and 3. The alternatives and the estimated job completion times in days are given in the table below.

	1	2	3
A	17	10	12
B	9	8	10
C	14	4	7

The manager wishes to minimise the total number of days required to complete all three jobs. How should the allocation of workers to jobs be made?

The table of numbers $\begin{pmatrix} 17 & 10 & 12 \\ 9 & 8 & 10 \\ 14 & 4 & 7 \end{pmatrix}$ is called the **cost matrix** for the problem.

The data can be modelled by the following network:

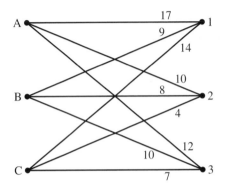

This is a bipartite graph (see Book D1) in which the nodes correspond to the workers and the jobs, and the arcs represent the possible assignments. The 'supply' at each source is 1 and the 'demand' at each destination is also 1. The 'cost' of assigning a worker, A, B or C, to a job, 1, 2 or 3, is the time taken for the worker to complete the job.

Comparing the above representation with the one in chapter 1 clearly shows the similarity of the network models of the assignment problem and the transportation problem. The assignment problem is a special case of the transportation problem in which:

(i) supply $= 1$
(ii) demand $= 1$
(iii) amount shipped over each arc is either 0 or 1.

2.2 Formulation as a linear programming problem

Because the assignment problem is a special case of the transportation problem, a linear programming formulation can be developed as we did in chapter 1.

We need a decision variable for each arc. For the arc from node i to node j we define:

$$x_{ij} = \begin{cases} 1 & \text{if worker } i \text{ does job } j \\ 0 & \text{otherwise} \end{cases}$$

We obtain the constraints by considering each node in turn:

(i) Since each worker is assigned to one and only one job we have:

for A $\quad x_{A1} + x_{A2} + x_{A3} = 1$ $\qquad$ (1a)

for B $\quad x_{B1} + x_{B2} + x_{B3} = 1$ $\qquad$ (1b)

for C $\quad x_{C1} + x_{C2} + x_{C3} = 1$ $\qquad$ (1c)

(ii) Since each job is assigned to one and only one worker we have:

for 1 $\quad x_{A1} + x_{B1} + x_{C1} = 1$ $\qquad$ (2a)

for 2 $\quad x_{A2} + x_{B2} + x_{C2} = 1$ $\qquad$ (2b)

for 3 $\quad x_{A3} + x_{B3} + x_{C3} = 1$ $\qquad$ (2c)

To obtain the objective function we consider completion times:

for A $\quad 17x_{A1} + 10x_{A2} + 12x_{A3}$ $\qquad$ (3a)

for B $\quad 9x_{B1} + 8x_{B2} + 10x_{B3}$ $\qquad$ (3b)

for C $\quad 14x_{C1} + 4x_{C2} + 7x_{C3}$ $\qquad$ (3c)

The sum of the three completion times gives us the total number of days needed to complete the three jobs. The objective is therefore to minimise:

$$Z = 17x_{A1} + 10x_{A2} + 12x_{A3}$$
$$+ 9x_{B1} + 8x_{B2} + 10x_{B3} \qquad (4)$$
$$+ 14x_{C1} + 4x_{C2} + 7x_{C3}$$

The linear programming problem is then: minimise Z, given by equation (4), subject to the set of constraints given by equations (1) and (2). The conditions (1) and (2) ensure that the variables are 0 or 1 and so no non-negativity condition is required.

Exercise 2A

1 Three machines, A, B and C, are available for producing three different components on a one-to-one basis. The time, in minutes, required by each machine to produce each component is given in the following table:

		Component		
		1	**2**	**3**
	A	13	14	10
Machine	**B**	11	16	15
	C	17	12	9

The objective is to assign machines to components so that the total time required is a minimum. Formulate this as a linear programming problem.

2.3 Solution of the assignment problem

Although the assignment problem is a special case of the transportation problem, it has a high degree of degeneracy. For a 4×4 cost matrix the magic number (see chapter 1) is $4 + 4 - 1 = 7$. However, as we have seen, the solution of the assignment problem requires only one cell in each row and one cell in each column to be occupied, that is only four occupied cells. This means that the transportation algorithm is a very inefficient way of solving the assignment problem.

The method described below was developed in 1955 by Harold W. Kuhn. He named it the **Hungarian algorithm** as it draws heavily on mathematical theories developed in 1916 by two Hungarians, D. König and E. Egevary.

The Hungarian algorithm

This algorithm depends on two features of the assignment problem:

(i) Each worker must be assigned to **one and only one job** and vice versa.
(ii) In the cost matrix a constant can be added or subtracted from all cost values in a row or all cost values in a column without having any effect on the set of optimal assignments.

■ **There are three steps in the Hungarian algorithm:**

Step 1 Find the opportunity cost matrix.
Step 2 Test for an optimal assignment. If an optimal assignment can be made, make it and stop.
Step 3 Revise the opportunity cost matrix and return to step 2.

The Hungarian algorithm is easier to apply than the transportation algorithm. It only requires some accurate addition and subtraction and a careful application of the test for optimality.

Finding the opportunity cost matrix

To obtain the opportunity cost matrix we start from the cost matrix for the problem. This will be similar to that given in Example 1. We then carry out two operations:
(i) row reduction – subtract the smallest number in each row from every number in that row
(ii) column reduction – then subtract the smallest number in each column from every number in that column.

Carrying out (i) will produce at least one zero in each row and carrying out (ii) will produce at least one zero in each column.

Subtracting the smallest number in row 1 from the other numbers in row 1 will produce the best course **for that row**, the zero, and the penalty or 'lost opportunity' for all the other values in that row, hence the name 'opportunity cost matrix' for the result of carrying out row and column reductions.

If the resulting opportunity cost matrix has one zero in each row and each column, then it is possible to make optimal assignments, that is assignments for which all of the opportunity costs are zero.

A simple test has been devised to help determine whether or not an optimal assignment can be made and this is considered in the next section.

Let us apply step 1 to Example 1. The cost matrix and row minima are shown below.

			Row minimum
17	10	12	10
9	8	10	8
14	4	7	4

After carrying out the row reduction we obtain:

	7	0	2
	1	0	2
	10	0	3
Column minimum	1	0	2

The column minima are shown above. The result of carrying out the column reduction is:

6	0	0*
0*	0	0
9	0*	1

It is easy to see by inspection that the starred (*) entries give an optimal assignment with one zero in each row and one zero in each column. Returning to the original cost matrix we have:

17	10	12*
9*	8	10
14	4*	7

The assignment is:

$$A \rightarrow 3 \quad \text{cost } 12$$
$$B \rightarrow 1 \quad \text{cost } 9$$
$$C \rightarrow 4 \quad \text{cost } 4$$

Total minimum cost $= 12 + 9 + 4 = 25$

Notice that this optimal assignment does not use the smallest number in row 1 or the smallest number in row 2.

Testing for an optimal assignment

■ **To test for an optimal assignment we draw the minimum number of straight lines (vertical and horizontal) on the opportunity cost matrix to cover all the zeros. Each line is drawn to cover as many zeros as possible at one time.**

If the number of lines is equal to the number of rows and columns in the matrix then an assignment can be made. If on the other hand the number of lines is less than the number of rows or columns then an optimal assignment cannot be made. In this case we must proceed to step 3 and develop a new opportunity cost matrix.

Example 1 (again)
The opportunity cost matrix we obtained above was:

6	0	0
0	0	0
9	0	1

A minimum of three lines is required to cover the zeros. One possible set of lines is shown above. Since this is equal to the number of rows and columns, an optimal assignment can be made. We will consider 'making the final assignment' after we have considered step 3 of the algorithm.

Example 2
Given the cost matrix:

25	4	15	7
6	3	8	18
13	2	2	4
1	1	2	1

find the opportunity cost matrix and test to see if an optimal assignment can be made.

The row minima are:

row 1 – 4
row 2 – 3
row 3 – 2
row 4 – 1

So reducing the rows we obtain:

21	0	11	3
3	0	5	15
11	0	0	2
0	0	1	0

The column minima are all zero and so column reduction produces no change. This is therefore the opportunity cost matrix. Drawing the minimum number of lines produces:

21	0	11	3
3	0	5	15
11	0	0	2
0	0	1	0

Since the minimum number of lines is three and the number of rows and columns is four **an optimal assignment cannot be made**.

Exercise 2B

1 The cost matrix for question 1 of Exercise 2A was:

13	14	10
11	16	15
17	12	9

(a) Obtain the opportunity cost matrix.

(b) Test this matrix to decide if an optimal assignment can be made.

(c) If this is possible, obtain such an assignment and its cost.

2 Three taxis, 1, 2 and 3, are available and there are three customers, A, B and C, requiring taxis. The distances between the taxis and the customers are shown in the table below, in kilometres. The company wishes to assign the taxis to customers so that the total distance travelled is a minimum.

Customers

	A	B	C
1	27	18	10
2	16	15	19
3	20	14	12

Taxis (rows 1, 2, 3)

(a) Obtain the opportunity cost matrix.

(b) Test this matrix to decide if an optimal assignment can be made.

3 The personnel director of a company must assign four recently qualified graduates to four regional offices. The graduates are all equally well qualified so the decision will be based on the costs of relocating the graduates' families. The cost data are presented in the following table, in units of £100.

Office

	N	S	E	W
Arthur	30	22	24	25
Baldwin	26	22	36	23
Coutts	43	21	23	39
Dinsdale	40	22	23	39

Graduate

(a) Obtain the opportunity cost matrix.

(b) Test this matrix to decide if an optimal assignment can be made.

Revising the opportunity cost matrix

An optimal assignment is rarely obtained from the initial opportunity cost matrix as we did in Example 1. It is usually necessary to revise the matrix in order to shift one, or more, of the zero costs from its present location (covered by lines) to a new uncovered location in the matrix.

Intuitively we would like this uncovered location to emerge with a zero opportunity cost.

■ **Revising the opportunity cost matrix is accomplished by carrying out two operations:**
  (i) **Subtract the smallest number not covered by a line from all numbers not covered by a straight line.**
  (ii) **Add this number to every number (including zeros) lying at the intersection of any two lines.**

Let us consider Example 2. The initial opportunity cost matrix obtained there was:

21	0	11	3
3	0	5	15
11	0	0	2
0	0	1	0

For the shown set of lines the smallest uncovered number is 3, so this value is subtracted from each of the six uncovered numbers. This value is then added to the two numbers that lie at the intersection of the lines.

The result is:

18	0	8	0
0	0	2	12
11	3	0	2
0	3	1	0

The minimum number of lines required to cover the zeros is now four. One such covering is:

18	0	8	0
0	0	2	12
11	3	0	2
0	3	1	0

Since the minimum number of lines is equal to the number of rows and columns, an optimal assignment can be made. How to make this assignment is discussed in the next section.

Making the final assignment

In making the final assignment you should, if possible, first consider rows and columns that have only one zero entry.

For Example 1 we obtained the final opportunity cost matrix:

6	0	0*
0*	0	0
9	0*	1

Column 1   has only one zero so we must use it (*).
Row 3   has only one zero so we must use it (*).

As row 2 and row 3, and also column 1 and column 2, have their cell designated, the remaining zero entry to be used must be (1, 3).

This agrees with the assignment made by inspection earlier. From the above we can see that it is the only optimal assignment.

Let us now return to the final opportunity cost matrix for Example 2:

18	0	8	0
0	0	2	12
11	3	0	2
0	3	1	0

Row 3 has only one zero entry and so an assignment can be made to that cell. Draw wavy lines through row 3 and column 3 where this cell occurs.

All the remaining rows and columns have two zeros.

Let us consider the two possible choices in row 1 and their consequences.
(i)   Choose (1, 2)
       ⇒ (1, 4) cannot be chosen so (4, 4) must be chosen
       ⇒ (2, 2) cannot be chosen so (2, 1) must be chosen
       The optimal assignment is then (3, 3), (1, 2), (4, 4) and (2, 1).
(ii)   Choose (1, 4)
       ⇒ (1, 2) cannot be chosen so (2, 2) must be chosen
       ⇒ (4, 4) cannot be chosen so (4, 1) must be chosen
       The optimal assignment is now (3, 3), (1, 4), (2, 2) and (4, 1).

If you return to the original cost matrix you may obtain the total minimum cost:

(i)   (1, 2) − 4, (2, 1) − 6, (3, 3) − 2, (4, 4) − 1
       Total 13
(ii)   (1, 4) − 7, (2, 2) − 3, (3, 3) − 2, (4, 1) − 1
       Total 13

In this case we have more than one optimal assignment but the cost of both of them is the same, 13. The problem is said to have **multiple optimal solutions**.

Exercise 2C

1 For question 2 of Exercise 2B the opportunity cost matrix is:

16	8	0
0	0	4
7	2	0

The zeros can be covered by two lines, as shown. Revise the opportunity cost matrix as often as necessary and hence obtain the optimal assignment for this problem and its cost.

2 In question 3(b) of Exercise 2B you should have shown that the opportunity cost matrix is:

4	0	1	2
0	0	13	0
18	0	1	17
14	0	0	16

Starting from this matrix and the set of lines shown, use the Hungarian algorithm to obtain an optimal assignment and give its cost.

3 The manager of the computer centre at a technical college has four programming jobs that she wishes to assign to four programmers. The estimated costs of assigning a particular programmer to a particular job are shown in the table below.

		Job			
		1	**2**	**3**	**4**
	Bob	10	5	18	11
Programmer	**Sue**	3	2	4	5
	Jim	18	9	17	15
	Amy	11	6	19	10

The assignment is to be made so that the total cost is a minimum. Use the Hungarian algorithm to obtain an optimal assignment and find its cost.

2.4 Unbalanced assignment problems

The solution procedure for assignment problems discussed so far requires that the number of rows in the cost matrix be equal to the number of columns. In practice this is often not the case.

The number of agents may exceed the number of tasks or vice versa. If there is one more agent than the number of tasks then there will be one more row than columns. We then simply add a **dummy column**, as we did in the transportation problem. If the number of columns is one more than the number of rows then we add a **dummy row**.

This then creates a table in which (number of rows) = (number of columns). Since the dummy agent or task is really non-existent, **the entries in the dummy row or column will all be zeros**. The addition of a dummy row or dummy column does not affect the solution method.

Example 3

A company has three employees and four machines and wishes to assign employees to machines to minimise total costs. The cost matrix showing the cost, in units of £10, incurred by each employee on each machine is given below.

	1	2	3	4
A	8	11	12	10
B	5	16	13	8
C	5	10	23	15

Determine the optimal assignment and calculate the total minimum cost.

As there are more machines than employees we add a dummy row D with zero entries and obtain the cost matrix:

	1	2	3	4
A	8	11	12	10
B	5	16	13	8
C	5	10	23	15
D	0	0	0	0

The row minima are 8, 5, 5 and 0, so reducing the rows we obtain the matrix:

0	3	4	2
0	11	8	3
0	5	18	10
0	0	0	0

As all the column minima are now zero, reduction of the columns does not change this matrix. The zeros may be covered by the two lines shown above.

The smallest uncovered number is 2. Adding and subtracting this from appropriate numbers, as specified earlier, we obtain the revised opportunity cost matrix:

0	1	2	0
0	9	6	1
0	3	16	8
2	0	0	0

The minimum number of lines required to cover the zeros is three, as shown above.

The smallest uncovered number is now 1. Adding and subtracting this from appropriate numbers we obtain the revised opportunity cost matrix:

0	0	1	0
0	8	5	1
0	2	15	8
3	0	0	1

Again a minimum number of three lines is required to cover the zeros, as shown above.

The smallest uncovered number is now 1 and proceeding as before we obtain:

1	0	1	0
0	7	4	0
0	1	14	7
4	0	0	1

Four lines are now required to cover the zeros, as shown, and so an optimal assignment can be made.

Column 3 has only one zero so use (4, 3).
If (4, 3) is used, then (4, 2) cannot be used $\Rightarrow$ use (1, 2).
If (1, 2) is used, then (1, 4) cannot be used $\Rightarrow$ use (2, 4).
The assignment is completed by using (3, 1).
The optimal assignment is then:

$$
\begin{array}{ll}
A \rightarrow 2 & \text{cost } 11 \\
B \rightarrow 4 & \text{cost } 8 \\
C \rightarrow 1 & \text{cost } 5 \\
D \rightarrow 3 & \text{cost } 0
\end{array}
$$

Total minimum cost $= 24 \times £10$.

The last assignment, $D \rightarrow 3$, means that no one is assigned to machine 3.

Exercise 2D

1 A book supplier has three salespersons to assign to four regions. The salespersons are able to cover the regions in different amounts of time. The amount of time, in days, required by each salesperson to cover each region is shown in the table below. Which salesperson should be assigned to which region in order to minimise total time? Obtain the optimal assignment by using the Hungarian algorithm and calculate the total time.

Regions

		A	B	C	D
Salesperson	1	10	2	8	6
	2	9	3	11	3
	3	3	1	4	2

2 Given the cost matrix:

	1	2
A	10	6
B	5	4
C	7	8

obtain two optimal assignments that have minimum total cost. Give this minimum total cost.

2.5 Maximisation assignment problems

Some assignment problems require the **maximisation** of profit or effectiveness rather than the minimisation of costs. It is straightforward to obtain an equivalent minimisation problem by converting all numbers in the given data matrix (called the payoff matrix) to opportunity costs. This is done by **subtracting every number in the original payoff matrix from the largest number in that matrix.** You may in fact subtract them from any number greater than or equal to the largest number. The arithmetic is often simplified by choosing a larger number, such as 100 or 1000. The transformed entries produced in this way are the opportunity costs.

We now apply the Hungarian algorithm to the transformed matrix. **The optimal assignment obtained is the optimal assignment for the original maximising problem.** The total payoff or profit is then obtained by adding the original payoffs of those cells used in the optimal assignment.

Example 4

A company has leased a new store and wishes to decide how to assign four departments to four locations to maximise total profits. The table below gives the individual profits in £s:

	1	2	3	4
Shoes	4	12	10	11
Toys	12	6	16	15
Hardware	16	20	18	16
Photography	13	16	15	14

The largest number in the matrix is 20. We will subtract all numbers from 20 as the arithmetic is easy. You may, of course, use your calculator to do or check these subtractions. The opportunity cost matrix obtained is then:

16	8	10	9
8	14	4	5
4	0	2	4
7	4	5	6

The row minima are 8, 4, 0 and 4.

Row reduction gives:

8	0	2	1
4	10	0	1
4	0	2	4
3	0	1	2

The column minima are 3, 0, 0 and 1.

Column reduction gives:

5	0	2	0
1	10	0	0
1	0	2	3
0	0	1	1

As shown above, the minimum number of lines required to cover the zeros is four and so an optimal assignment can be made.

There is only one zero in column 1 so use (4, 1).
There is only one zero in column 3 so use (2, 3).
There is only one zero in row 3 so use (3, 2).

Using (2, 3) $\Rightarrow$ do not use (2, 4), therefore use (1, 4) to complete the assignment.

It is suggested that you complete the table below as you proceed.

	0		0*
		0*	0
	0*		
0*	0		

Returning to the original matrix we obtain:

(1, 4): shoes $\Rightarrow$ 4, profit £11
(2, 3): toys $\Rightarrow$ 3, profit £16
(3, 2): hardware $\Rightarrow$ 2, profit £20
(4, 1): photography $\Rightarrow$ 1, profit £13

Total profit $=$ £60

Exercise 2E

1 A head of department has four teachers to be assigned to four
   different courses. All of the teachers have taught the courses in
   the past and have been evaluated by the students. The rating
   for each teacher for each course is given in the table below, a
   perfect score is 100. The head of department wants to know
   the optimal assignment of teachers to courses that will
   maximise the overall average evaluation.

	A	B	C	D
1	85	80	80	75
2	75	85	78	81
3	83	81	70	74
4	81	82	83	78

Use the Hungarian algorithm to solve this assignment problem.

SUMMARY OF KEY POINTS

1 The Hungarian algorithm

   Step 1 Find the opportunity cost matrix.
   Step 2 Test for an optimal assignment. If an optimal
   assignment can be made, make it and stop.
   Step 3 Revise the opportunity cost matrix and return to
   step 2.

2 Testing for an optimal assignment

   If the minimum number of straight lines (vertical and
   horizontal) required to cover all the zeros, in the
   opportunity cost matrix, is equal to the number of rows
   and columns in the matrix then an assignment can be
   made.

3 Revising the opportunity cost matrix

   To revise the opportunity cost matrix
   (i) **Subtract** the smallest number not covered by a line
   from all numbers not covered by a straight line.
   (ii) **Add** this number to every number (including zeros)
   lying at the intersection of any two lines.

T.S.P. $\longrightarrow$ classical = visit each vertex only once

Practical = visit each vertex at least once

$$T_p \leq T_c$$

The travelling salesman problem

3

3.1 What is the travelling salesman problem (T.S.P.)?

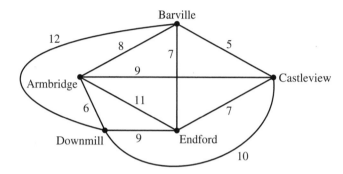

The diagram above shows five villages and the roads connecting them. The numbers indicate the distances, in miles, between the villages.

In chapter 4 of Book D1 we considered the problem of inspecting the roads, the so-called **route inspection problem**. We were looking for a route through the network that traversed each road and returned to the starting point and was of minimum length.

In this chapter we consider the analogous problem of finding a route that visits each village once and returns to its starting point and involves travelling a minimum distance. This is the problem encountered by a salesman who lives in Armbridge and wishes to visit customers in each of the other villages. For this reason it is usually known as the **travelling salesman problem** (T.S.P. for short).

The situation can be modelled by a network in which the vertices represent the villages, the edges represent the roads and the weights on the edges represent the distances. We then have the network shown on the right.

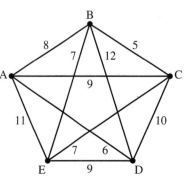

A **walk** in a network is a finite sequence of edges such that the end vertex of one edge is the start vertex of the next.

It is convenient to make the following definition to save a lot of writing:

■ **A tour is a route that visits every vertex in the network, returning to the starting vertex.**

Some possible tours in the above network are:
(i)   ABCDEA of length 43 miles
     (Notice that this is essentially the same as AEDCBA or
     BCDEAB or BAEDCB, etc. Depending on the choice of
     starting vertex and the direction taken there are 10 ways of
     writing it down.)

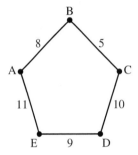

(ii) ACEBDA of length 41

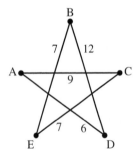

(iii) ABCEDA of length 35

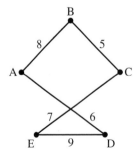

The travelling salesman problem is very easily stated using the above definition: **find a tour of minimum weight**.

In real life the salesman would not mind backtracking through a previously visited village, if this produced the shortest route. This problem is called the **practical T.S.P.** Here each vertex must be visited **at least once**.

If we impose the constraint that the tour must visit each vertex **exactly once** then we obtain the **classical T.S.P.**

Since a practical problem can be transformed into a corresponding classical problem, we will restrict our attention to the classical T.S.P. (This transformation is considered in section 3.6.)

The classical T.S.P. problem is:

■ **Find a tour of minimum weight that visits each vertex once only.**

This problem was first studied, in the 19th century, by Sir William Hamilton. In chapter 2 of Book D1 we defined a cycle that includes all vertices of a network as a **Hamiltonian cycle**. The T.S.P. problem is then to **find a Hamiltonian cycle of least weight**.

Not all networks have Hamiltonian cycles but it can be shown that all **complete graphs** K_n ($n \geqslant 3$) do have Hamiltonian cycles. We will therefore restrict our attention, initially, to complete graphs.

> A **complete graph** is one in which each vertex is joined to every other vertex.

In addition we will assume that the network satisfies the triangle inequality:

■ **For all sets of vertices {A, B, C},
weight BC $\leqslant$ weight AB + weight AC.**

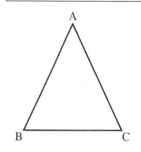

3.2 The difficulty with the T.S.P.

Although the T.S.P. is easy to state, there is no known algorithm that provides a quick and efficient solution.

A complete enumeration algorithm is one that systematically tries every possibility. Let us consider such an algorithm for the classical T.S.P. The steps are:

Step 1   List all Hamiltonian cycles.
Step 2   Find the total weight of each cycle.
Step 3   Choose the cycle of least total weight. (There may be more than one such cycle.)

For our earlier problem, with five vertices there are $4 \times 3 \times 2 \times 1 = 24$ cycles to consider, namely

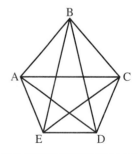

ABCDEA; ABCEDA; ABDCEA; ABDECA; ABECDA;
ABEDCA; ACBDEA; ACBEDA; ACDBEA; ACDEBA;
ACEBDA; ACEDBA; ADBCEA; ADBECA; ADCBEA;
ADCEBA; ADEBCA; ADECBA; AEBCDA; AEBDCA;
AECBDA; AECDBA; AEDBCA; AEDCBA.

> $4 \times 3 \times 2 \times 1$ is usually written 4! and called **4 factorial**.

In the general case, when we have n vertices the number of cycles is $n!$ and when $n = 100$ this is of order 10^{155}, which is way beyond present computer capacities.

As there is no efficient algorithm for the T.S.P. we are forced to look for approximate solutions, that is we look for upper and lower bounds for the weight of a minimum weight Hamiltonian cycle.

If we can show that the best solution will have weight **at most** x then x is an **upper bound**.

If we can show that the best solution will have weight **at least** y then y is a **lower bound**.

The **smaller** we can make the **upper bound**, the more useful it will be and the **larger** we can make the **lower bound**, the more useful it will be. If x and y are close together and we can find a tour with total weight between x and y then we will be satisfied with that tour, even though we are not sure it is the best tour.

In the next two sections we will consider methods for finding upper and lower bounds using minimum spanning trees. You will recall that minimum spanning trees are discussed in chapter 3 of Book D1, where Kruskal's and Prim's algorithms are given.

3.3 Finding an upper bound

It can be shown that for a complete network that satisfies the triangle inequality an upper bound to the length of an optimal tour is:

■   **2 × (weight of a minimum spanning tree)**

This corresponds to going along each edge of the minimum spanning tree in both directions. The route so formed visits each vertex twice.

A typical example starting and finishing at L is shown below.

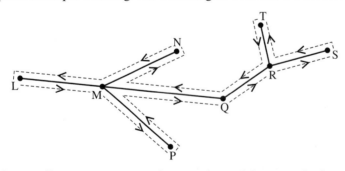

This is usually not a very good upper bound but can be improved by making shortcuts, that is, for example, replacing TRQMN by TN.

When making shortcuts do make sure that you end up with a route that still visits all vertices at least once.

This technique is illustrated in the examples below.

Example 1

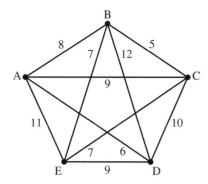

A minimum spanning tree for this network is easily obtained by using either Kruskal's algorithm or Prim's algorithm (see Book D1 chapter 3, sections 3.2, 3.3 and 3.4).

There are two possible minimum spanning trees:

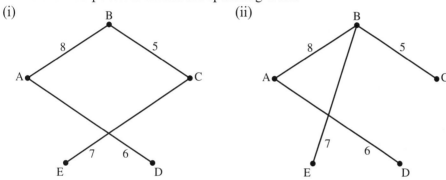

(i) Total length 26 units

(ii) Total length 26 units

According to the result quoted above, **twice the length of a minimum spanning tree** is an upper bound for the total length of the optimal solution.

So an upper bound is $2 \times 26 = 52$ units.

Let us concentrate for a moment on (i). In order to obtain shortcuts you are advised to draw the minimum spanning tree so that no edges cross. For example, we draw (i) as:

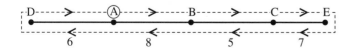

The route of length 52 units ABCECBADA starting and finishing at A is shown above by the dotted line.

However, when we reach E we could move directly, without visiting C, B and A again, to D, that is replacing ECBAD by ED. This is a shortcut.

When this shortcut is made the route is:

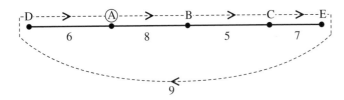

Since ED is of length 9, and $EC + CB + BA + AD = 7 + 5 + 8 + 6 = 26$, this produces a saving of $26 - 9 = 17$. We now have a better upper bound of $52 - 17 = 35$ units.

Let us look now at (ii). This minimum spanning tree may be drawn as:

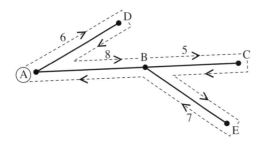

The route of length 52 units starting at A is now ADABCBEBA, again shown by a dotted line.

Two possible shortcuts are:
(i) directly from D to C instead of DABC; this gives a saving of $(8 + 6 + 5) - 10 = 9$
(ii) directly from C to E instead of CBE; this is a saving of $(7 + 5) - 7 = 5$.

If we use **both** shortcuts the saving is $9 + 5 = 14$ and the upper bound is now $52 - 14 = 38$.

When both shortcuts are used the route is:

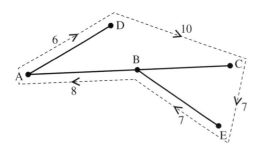

When we have made the shortcuts the resulting route does still visit each of the vertices.

The better of the two improved upper bounds we have produced is 35. We can then say that:

(total length of optimal solution) $\leqslant$ 35 units.

Example 2

	L	**C**	**O**	**B**	**N**	**E**
London (L)	—	80	56	120	131	200
Cambridge (C)	80	—	100	98	87	250
Oxford (O)	56	100	—	68	103	154
Birmingham (B)	120	98	68	—	54	161
Nottingham (N)	131	87	103	54	—	209
Exeter (E)	200	250	154	161	209	—

A sales representative, Sheila, has to visit clients in six cities: London, Cambridge, Oxford, Birmingham, Nottingham and Exeter. The table shows the distances, in miles, between these six cities. Sheila lives in London and plans a route starting and finishing in London. She wishes to visit each city and drive the minimum distance.

(a) Starting from London, use Prim's algorithm to obtain a minimum spanning tree.
(b) Hence determine an initial upper bound for the length of the route planned by Sheila.
(c) Starting from your initial upper bound and using shortcuts, obtain a route that is less than 660 miles. Can you improve on this?

	1	5	2	3	4	6
	L	**C**	**O**	**B**	**N**	**E**
L	—	80	56	120	131	200
C	⑧⓪	—	100	98	87	250
O	⑤⑥	100	—	68	103	154
B	120	98	⑥⑧	—	54	161
N	131	87	103	⑤④	—	209
E	200	250	①⑤④	161	209	—

Carrying out Prim's algorithm as shown in the above table (see Book D1 chapter 3, section 3.4) we obtain the following tree:

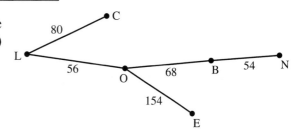

The arcs were selected in the order

LO (56), OB (68), BN (54), LC (80) and OE (154).

(b) The total weight of the minimum spanning tree is $(56 + 68 + 54 + 80 + 154) = 412$. Hence an initial upper bound for the length of the route planned by Sheila is $2 \times 412 = 824$ miles.

(c) If we use the shortcut NC to replace NBOLC we have:

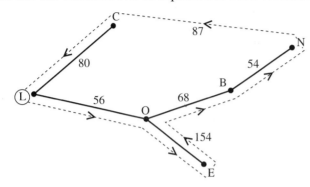

The length of this route is

$$56 + 154 + 154 + 68 + 54 + 87 + 80 = 653 \text{ miles}$$

We could also have obtained this result by considering the saving, which is

$$(54 + 68 + 56 + 80) - 87 = 171$$

This gives an upper bound of $824 - 171 = 653$ miles.

This upper bound can, however, be further improved by using the additional shortcut EB to replace EOB. This produces a further saving of $(154 + 68) - 161 = 61$. This reduces the upper bound to $653 - 61 = 592$ miles.

The route is now LOEBNCL. This is in fact a tour visiting each vertex only once.

Exercise 3A

1

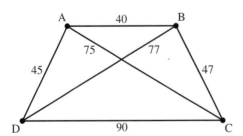

(a) Find a minimum spanning tree for the above network and hence find an upper bound for the total length of the solution of the travelling salesman problem.

(b) Obtain a better upper bound by using a shortcut.

2

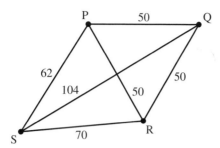

For the above network obtain three minimum spanning trees.
Hence obtain an upper bound for the total length of the
travelling salesman problem. By using shortcuts obtain an
upper bound that is less than 240.

3

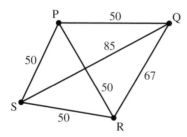

(a) Obtain three minimum spanning trees for the above
network and hence obtain an upper bound for the total length
of the solution of the travelling salesman problem.
(b) Consider each of these trees in turn and obtain in each
case a better bound by using shortcuts.

4

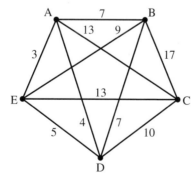

A delivery van based at A is required to deliver goods in
towns B, C, D and E shown in the diagram. The numbers on
the arcs are distances in miles.
(a) Find two minimum spanning trees for the above network
and hence find an upper bound for the distance travelled by
the van.
(b) Improve this upper bound by using shortcuts.

5 A computer engineer lives in town A and needs to visit each of the towns B, C, D and E to service various installations. He must return to town A and visit each town once. The distances in miles between the towns are shown in the table.

	A	B	C	D	E
A	—	17	10	9	12
B	17	—	8	14	5
C	10	8	—	7	11
D	9	14	7	—	11
E	12	5	11	11	—

(a) Use Prim's algorithm to find the length of a minimum spanning tree that connects the five towns.

(b) Hence find an upper bound for the total distance that must be travelled by the engineer.

(c) Improve the upper bound by using shortcuts.

6 The table shows the distances, in miles, between some cities. A politician has to visit each city once, starting and finishing at A. She wishes to minimise her total travelling distance.

	A	B	C	D	E	F	G	H
A	—	47	84	382	120	172	299	144
B	47	—	121	402	155	193	319	165
C	84	121	—	456	200	246	373	218
D	382	402	456	—	413	220	155	289
E	120	155	200	413	—	204	286	131
F	172	193	246	220	204	—	144	70
G	299	319	373	155	286	144	—	160
H	144	165	218	289	131	70	160	—

(a) Find a minimum spanning tree for the network.

(b) Hence find an upper bound for the politician's problem.

(c) Reduce this upper bound to a value below 1400 miles by using shortcuts.

3.4 Finding lower bounds

Suppose that in a network a minimum length Hamiltonian cycle has been found. If we delete a vertex A, say, and its incident edges then we get a path passing through the remaining vertices. This path must be a spanning tree for the graph formed by the remaining vertices but it may not be a minimum spanning tree:

∴   (sum of weight of edges on path)
   ≥ (weight of minimum spanning tree of reduced
      network)   (1)

When the vertex A and its incident edges are deleted, two of the edges, AU and AV say, of the Hamiltonian cycle are deleted. The sum of the weights of AU and AV must be at least equal to the sum of the weights of the two shortest edges at A:

∴   (weight AU + weight AV)
   ≥ (sum of the weights of the two shortest edges
      incident at A)   (2)

■   **Taken together equations (1) and (2) give us the following recipe for a lower bound:**

   (weight of the minimum spanning tree of the graph obtained by deleting a vertex V) + (length of the two shortest edges incident at V)

Based on this result we have the following algorithm.

Step 1   Choose any vertex V and delete it and all edges that are connected to it.

Step 2   Use a minimum connector algorithm to find the length of a minimum connector (minimum spanning tree) for the remaining vertices.

Step 3   Add to that length the sum of the lengths of the two shortest edges incident at V.

A set of lower bounds can be constructed by deleting each vertex in turn. The largest of the lower bounds is then chosen.

Step 4   Repeat steps 1 to 3, choosing each vertex in turn. Choose the largest result.

Notice that this method of producing a lower bound often does not produce a tour.

Example 3

Find lower bounds for the network considered in Example 1.
For ease of reference we repeat the network opposite.
Deleting A and the edges connected to A gives:

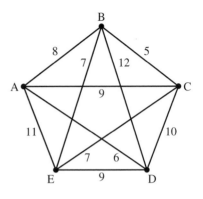

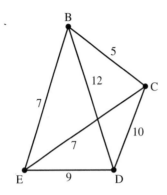

There are two possible minimum spanning trees:

(i)

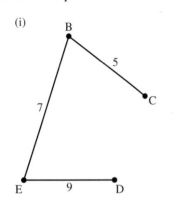

(ii)

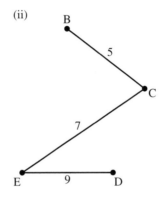

Each of these has weight 21.
The two edges of shortest length that were deleted are AD (6) and
AB (8).

Adding these in gives:

(i)

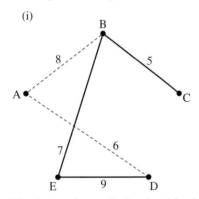

(ii)

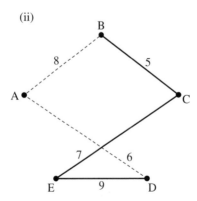

The lower bound obtained is then:

$$21 + 6 + 8 = 35$$

Notice that in case (ii) addition of the edges AB and AD does in fact produce a tour.

If we repeat this process for the other vertices we obtain the following results:

Deleted vertex	A	B	C	D	E
Lower bound	21 + (6 + 8) = 35	22 + (7 + 5) = 34	21 + (7 + 5) = 33	20 + (9 + 6) = 35	19 + (7 + 7) = 33

The best lower bound is the largest of these, namely 35. We can then say:

$$\text{(total length of optimal solution)} \geqslant 35 \text{ units}$$

If we combine this with the result obtained from our upper bounds in Example 1 we obtain:

$$35 \leqslant \text{(total length of optimal solution)} \leqslant 35$$

In this example we therefore have (lower bound) = (upper bound.)

Thus we have a solution, namely ABCEDA.

Example 4

Find lower bounds for the problem given by the distance matrix in Example 2. For ease of reference we repeat the matrix below.

	L	C	O	B	N	E
L	—	80	56	120	131	200
C	80	—	100	98	87	250
O	56	100	—	68	103	154
B	120	98	68	—	54	161
N	131	87	103	54	—	209
E	200	250	154	161	209	—

As it will be useful in what follows, we also repeat below the minimum spanning tree and the order in which arcs were selected.

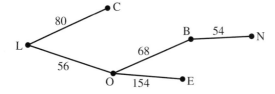

Order: LO (56), OB (68), BN (54), LC (80), OE (154)

(i) Deleting C and the arcs connected to C.

We could proceed to find the minimum spanning tree of the resulting network by deleting the row and column labelled C. However, by looking at the above tree we see that C occurs at the end of a branch and so the minimum spanning tree can be obtained merely by removing the arc LC, giving:

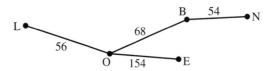

The two arcs of shortest length that were deleted are easily identified by looking along the C row. They are CL (80) and CN (87). Adding these in gives:

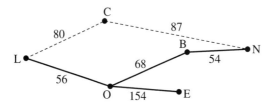

The lower bound obtained is then:

(weight of minimum spanning tree) + (weight CL + weight CN)
  = (56 + 68 + 54 + 154) + (80 + 87)
  = 332 + 167
  = 499 miles

The results of deleting E and N in turn may be obtained in a similar manner as they also occur at the end of a branch in the minimum spanning tree.

(ii) Deleting E and the arcs connected to E.

The minimum spanning tree and the connecting arcs are shown below.

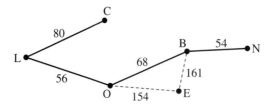

The lower bound is now:

$$(80 + 56 + 68 + 54) + (154 + 161)$$
$$= \quad 258 + (315)$$
$$= \quad 573 \text{ miles}$$

(iii) Deleting N and the arcs connected to N.
The minimum spanning tree and the connecting arcs are shown below.

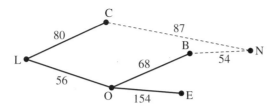

The lower bound obtained is now:

$$(80 + 56 + 68 + 154) + (87 + 54)$$
$$= 358 + 141$$
$$= 499 \text{ miles}$$

As neither L nor O occur at the end of a branch, we must return to the distance matrix and delete the appropriate rows and columns.
(iv) Deleting L and the arcs connected to L produces the reduced distance matrix:

		1	4	3	2	5
		C	**O**	**B**	**N**	**E**
C		—	100	98	87	250
O		100	—	(68)	103	154
B		98	68	—	(54)	161
N		(87)	103	54	—	209
E		250	(154)	161	209	—

Applying Prim's algorithm as shown in the above table we obtain the following tree:

The arcs were selected in the order CN (87), NB (54), BO (68), OE (154).

The two arcs of shortest length that were deleted are LO (56) and LC (80). These are shown in the diagram below.

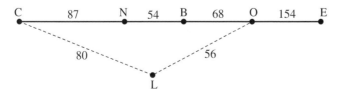

The lower bound is then:

$$(87 + 54 + 68 + 154) + (80 + 56)$$
$$= 363 + 136$$
$$= 499 \text{ miles}$$

(v) Deleting the rows and columns of the distance matrix labelled by O and proceeding in the same way we obtain the following tree:

The two arcs of shortest length that were deleted are OL (56) and OB (68). Adding these in gives

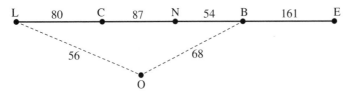

The lower bound obtained is now:

$$(80 + 87 + 54 + 161) + (56 + 68)$$
$$= 382 + 124$$
$$= 506 \text{ miles}$$

The above calculations of lower bounds are summarised in the table below.

Deleted vertex	**C**	**E**	**N**	**L**	**O**
Lower bound	499	573	499	499	506

The best lower bound is the largest of these, namely 573.

Taken together with the calculation of the upper bound carried out in Example 2 we can deduce:

$$573 \text{ miles} \leqslant \text{ length of optimal tour} \leqslant 592 \text{ miles}$$

In fact the route LOEBNCL obtained in the calculation of the upper bound is the optimal tour for this problem. It is of interest to see that this route can also be obtained from the figure given in (ii) above:

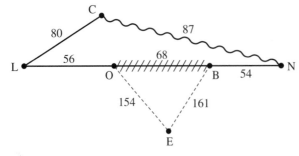

If we delete OB and add in CN we get the above route. Its length is $573 + 87 - 68 = 592$ miles.

Exercise 3B

1 In question 1 of Exercise 3A the following network was considered:

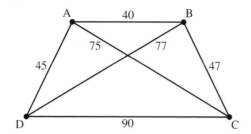

Obtain lower bounds for the length of the travelling salesman problem:

(a) by deleting A

(b) by deleting B

(c) by deleting C

(d) by deleting D.

2 In question 2 of Exercise 3A the following was considered:

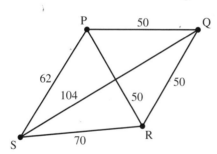

Obtain lower bounds for the length of the travelling salesman problem:

(a) by deleting Q

(b) by deleting S.

(c) Use your results for (a) and (b), together with your answers to question 2 of Exercise 3A, to obtain the tour in this network, of minimum weight.

3 In question 4 of Exercise 3A you had the following network:

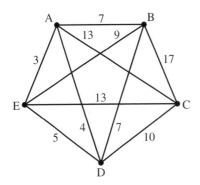

Obtain lower bounds for the length of the travelling salesman problem:

(a) by deleting A   (b) by deleting B   (c) by deleting C

(d) by deleting D   (e) by deleting E.

(f) Using your results for (a)–(e) and your answer to question 4 of Exercise 3A, write down an inequality for L, the minimum length of a tour.

(g) Write down a tour that satisfies this inequality.

4 In question 5 of Exercise 3A a matrix giving the distances between towns A, B, C, D and E was considered:

	A	B	C	D	E
A	—	17	10	9	12
B	17	—	8	14	5
C	10	8	—	7	11
D	9	14	7	—	11
E	12	5	11	11	—

Obtain lower bounds to the distance travelled by the computer engineer by:

(a) deleting A   (b) deleting B   (c) deleting C

(d) deleting D   (e) deleting E.

(f) From your answer to (e) and your answer to question 5 of Exercise 3A, obtain the minimum tour of length 41.

5 Below is show the distance, in miles, between some cities as used in question 6 of Exercise 3A.

	A	B	C	D	E	F	G	H
A	—	47	84	382	120	172	299	144
B	47	—	121	402	155	193	319	165
C	84	121	—	456	200	246	373	218
D	382	402	456	—	413	220	155	289
E	120	155	200	413	—	204	286	131
F	172	193	246	220	204	—	144	70
G	299	319	373	155	286	144	—	160
H	144	165	218	289	131	70	160	—

Use the minimum spanning tree that you found in question 6 of Exercise 3A to obtain lower bounds to the distance to be travelled by the politician:

(a) by deleting B   (b) by deleting C   (c) by deleting D.

(d) Write down an inequality satisfied by L, the minimum length of the tour.

6 A lorry driver has to deliver milk every day to village shops. He starts and finishes at A and must visit villages B, C, D, E and F. The distances between the villages are shown in the table.

	A	B	C	D	E	F
A	—	11	13	8	15	13
B	11	—	10	16	5	6
C	13	10	—	17	8	8
D	8	16	17	—	17	16
E	15	5	8	17	—	4
F	13	6	8	16	4	—

(a) Obtain an upper bound for the distance the lorry must travel.

(b) Obtain a lower bound for the distance travelled.

(c) Using the best upper bound and lower bound you can obtain, write down an inequality for L, the minimum distance travelled.

(d) Write down the shortest possible route.

3.5 The nearest neighbour algorithm

The method for finding upper bounds discussed in section 3.3, the 'twice the minimum connector plus shortcuts' approach, can produce good upper bounds. However, in a large network it is often difficult to find a good set of shortcuts. Furthermore, it is difficult to make this method algorithmic.

One alternative approach is to use the **nearest neighbour algorithm**. This is a **heuristic algorithm** like the bin-packing algorithms discussed in chapter 1 of Book D1. A heuristic algorithm produces good results but is **not guaranteed** to obtain the best (optimal) result.

■ **The nearest neighbour algorithm is:**

Step 1   **Let V be the current vertex.**
Step 2   **Find the nearest unvisited vertex to the current vertex, move directly to that vertex and call it the current vertex.**
Step 3   **Repeat step 2 until all vertices have been visited and then return directly to the start vertex.**

We may choose to start the algorithm at any vertex. The tour produced is dependent on the choice of starting vertex. If we repeat the algorithm, choosing each vertex in turn as the starting vertex, then we can take the length of the shortest tour as our upper bound.

> **Warning** – a common error
>
> The nearest neighbour algorithm for the T.S.P. and Prim's algorithm for finding a minimum connector are often confused. In the nearest neighbour algorithm the next vertex is the nearest unvisited neighbour to the **current vertex**. In Prim's algorithm the next vertex is the nearest unconnected vertex to the **current connected set of vertices**.

The major drawback of this algorithm is that we have no control over the choice of the final edge – it may be very long. In addition, in applying the algorithm we may miss very short edges.

Example 5

Use the nearest neighbour algorithm to obtain an upper bound for the length of the optimal tour for the network opposite (this is the network given in Example 1).

If we start at A then we choose D then E. There are then two possibilities: B and C.

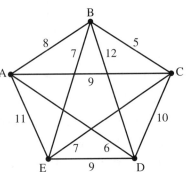

(i)   If we choose B then we choose C and complete the tour with edge CA. The tour is then ADEBCA of length $6 + 9 + 7 + 5 + 9 = 36$.

(ii) If we choose C then we choose B and complete the tour with BA. The tour is then ADECBA of length $6 + 9 + 7 + 5 + 8 = 35$.

The results of starting with the other vertices are summarised in the table below.

Starting vertex	Nearest neighbour tour(s)	Length
A	ADEBCA	36
	ADECBA	35
B	BCEDAB	35
C	CBEDAC	36
D	DABCED	35
E	EBCADE	36
	ECBADE	35

The best upper bound obtained from this is 35 units, as we found in Example 1.

Example 6

Use the nearest neighbour algorithm to find a tour, starting at London, for the data given in Example 2, which is reproduced below. Obtain a better upper bound by starting at another vertex.

	①L	⑤C	②O	③B	④N	E
L	—	80	56	120	131	200
C	80	—	100	98	㊼87	250
O	㊄56	100	—	68	103	154
B	120	98	㊳68	—	54	161
N	131	87	103	㊊54	—	209
E	200	㉕250	154	161	209	—

We may carry out the algorithm by examining entries in columns in a similar way to what we do when we apply Prim, bearing in mind the warning given earlier. We start the algorithm by choosing L, as the tour must start at London. Cross out row L and label column L with ①. Choose the smallest number in column L and circle it, ㊄56 (arc LO).

Cross out row O and label column O with ②.

Choose the smallest remaining number in column O and circle it, ㊳68 (arc OB).

Cross out row B and label column B with ③.
Choose the smallest remaining number in column B and circle it, (54) (arc BN).

Cross out row N and label column N with ④.
Choose the smallest remaining number in column N and circle it, (87) (arc NC).

Cross out row C and label column C with ⑤.
Choose the smallest remaining number in column C and circle it, (250) (arc CE).

Finally join E to L (arc EL), this is (200).
The tour we have is:

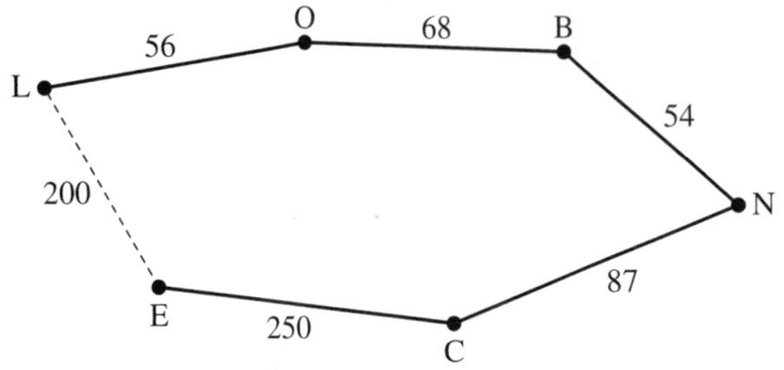

The total length of this tour is:

$$(56 + 68 + 54 + 87 + 250) + 200 = 715 \text{ miles}$$

This is an upper bound for the length of the optimal tour, which should be compared with those obtained earlier in this chapter.

It is not really a very good upper bound as the tour uses two long arcs to the vertex E. The above illustrates the weakness of this algorithm, that is we have no control over the choice of the final arc that completes the tour.

A shorter tour can be found by choosing a different starting vertex, chosen so that, for example, the shortest arc from E is used.

Using the nearest neighbour algorithm starting at E produces the following tour of length 592 (this is the optimal tour):

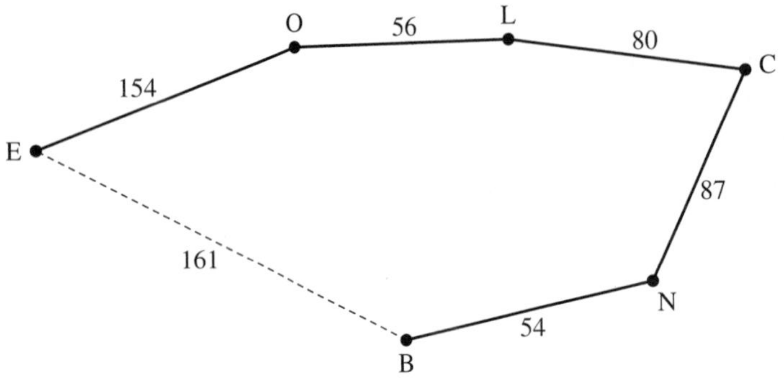

Exercise 3C

1

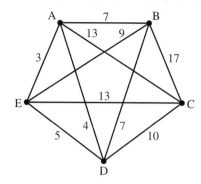

Use the nearest neighbour algorithm to obtain upper bounds for the length of the optimal tour for the network given above:

(a) starting at A

(b) starting at B.

State the tour in each case and give its length.

2

	A	B	C	D	E
A	—	17	10	9	12
B	17	—	8	14	5
C	10	8	—	7	11
D	9	14	7	—	11
E	12	5	11	11	—

Use the nearest neighbour algorithm to find a tour, starting at A, for the network given by the above matrix.

3.6 The practical T.S.P.

You can transform a practical T.S.P. into the corresponding classical problem by creating a complete network in which the edge weights are the shortest distances in the original network.

So far in this chapter we have considered only T.S.P.s for networks that are **complete** and for which the **triangle inequality holds** (see Book D1 chapter 2 for the definitions of these terms). In these networks the classical problem and the practical problem are the same.

The complete network, whose creation is referred to above, will by construction also satisfy the triangle inequality. We can now therefore use the methods discussed earlier in this chapter.

Solving the classical problem in the transformed network is equivalent to solving the practical problem in the original network. The method is illustrated in the following two examples.

Example 7

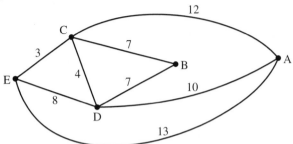

The network shows a number of villages, A, B, C, D and E, in the area covered by a baker's delivery van and the roads joining them. The number on the edge gives the distance between the villages, in kilometres.

(a) Draw a complete network, showing the shortest distances between the villages.

(b) Use the nearest neighbour algorithm on the complete network, drawn in (a), to obtain an upper bound to the length of a tour in this network that starts at A and visits each village exactly once.

(c) Interpret your result in (b) in terms of the original network.

(a) The given network is clearly **not complete** since, for example, B is not connected directly to A or to E. It also does not satisfy the triangle inequality since in triangle CDE we have

$$CD + CE = 4 + 3 = 7$$

and this is less than DE, which is 8.

To make the graph complete we must add edges AB and BE. The shortest distances are:

$$AB: \quad AD + DB = 10 + 7 = 17 \, \text{km}$$
$$BE: \quad BC + CE = 7 + 3 = 10 \, \text{km}$$

These have been written down by inspection. In more complicated cases you may need to use Dijkstra's algorithm (see Book D1 chapter 3, section 3.5).

The edges in the given figure are all the shortest routes between pairs of villages, except for edge DE. The shortest route from D to E is $DC + CE = 4 + 3 = 7$.

The complete network showing the shortest distances between villages is then:

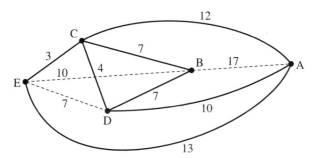

(b) Using the nearest neighbour algorithm starting at A, on the network obtained in (a), gives:

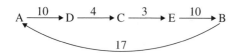

The tour is ADCEBA and is of length 44 miles.

(c) Returning to the original network we see that EB must be replaced by EC and CB, since EB is not in the original network. Similarly, BA must be replaced by BD and DA.

The route in the original network is then ADCECBDA. This is, of course, of length 44 miles and the tour visits C and D twice.

Example 8

(a) For the network opposite, draw up a table in which the entries are the shortest distances between pairs of vertices.

(b) Use the nearest neighbour algorithm on this table to obtain an upper bound to the length of the tour, in the complete network, which starts and finishes at A and visits every vertex exactly once.

(c) Interpret your answer in (b) in terms of the original network.

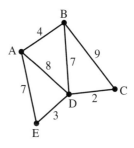

(a)

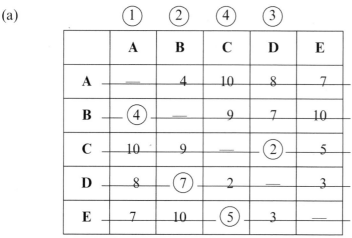

	A	B	C	D	E
A	—	④ 4	10	8	7
B	4	—	9	7	10
C	10	9	—	② 2	5
D	8	⑦ 7	2	—	3
E	7	10	⑤ 5	3	—

The arcs shown in the given network are all the shortest distances between relevant vertices.

The shortest route from A to C is ADC, of length 10 [alternatives are ABC (13), ABDC (13) and AEDC (12)].

The shortest route from B to E is BDE, of length 10 [alternatives are BAE (11) and BCDE (14)].

The shortest route from C to E is CDE (5).

It is suggested that even when you are finding shortest routes by inspection you list all possible routes.

(b) Using the nearest neighbour algorithm on the table gives:

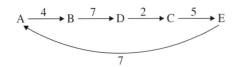

The upper bound is $4 + 7 + 2 + 5 + 7 = 25$. The tour is ABDCEA.

(c) Since CE does not occur in the original network, the tour in the original network is ABDCDEA with CD and DE replacing CE.

Exercise 3D

1

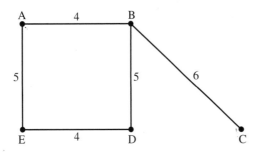

(a) For the network shown above, draw a complete network in which the arc lengths are the shortest distances between vertices.

(b) Use the nearest neighbour algorithm, starting at A, to find a tour that visits each vertex exactly once. Give its length.

(c) Interpret your result in part (a) in terms of the original network.

2

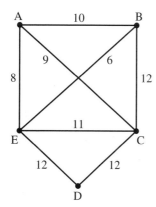

(a) For the above network draw up a table in which the entries are the shortest distances between pairs of vertices.

(b) Use the nearest neighbour algorithm, starting at A, to obtain a tour in the transformed network. Give its length and interpret the tour in terms of the original network.

(c) Use the nearest neighbour algorithm, starting at different vertices, to obtain tours that have a shorter length. Interpret these tours in terms of the original network.

3

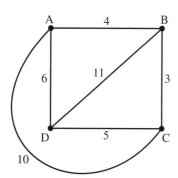

For the complete graph shown above draw up a table in which the entries are the shortest distances between pairs of vertices.

SUMMARY OF KEY POINTS

1 A **tour** is a route that visits every vertex in the network returning to the starting vertex.

2 An **upper bound** to the length of an optimal tour is twice the weight of a minimum spanning tree.

3 A **lower bound** to the length of an optimal tour is given by: weight of the minimum spanning tree obtained by deleting a vertex V + length of the two shortest edges incident at V.

4 The nearest neighbour algorithm for obtaining an upper bound is:

Step 1 Let V be the current vertex.

Step 2 Find the nearest unvisited vertex to the current vertex, move **directly** to that vertex and call it the current vertex.

Step 3 Repeat step 2 until all vertices have been visited and then return **directly** to the start vertex.

Game theory

4

4.1 Basic definitions

There are many situations in which several decision makers compete with one another to arrive at the best outcome. These competitive decision-making situations are the subject of **game theory**. The study of game theory dates back to 1944, the year in which John von Neumann and Oskar Morgenstern published their classic book *Theory of Games and Economic Behaviour*.

If you have played card games or board games you will be familiar with situations in which competitors develop plans of action in order to win. Game theory deals with similar situations in which competing decision makers develop plans of action in order to win. In addition, game theory develops several mathematical techniques to aid the decision maker in selecting the plan of action that will result in the best outcome.

4.2 Types of game situations

Competitive games can be subdivided into several categories. One classification is based on the number of competitive decision makers, called **players**, involved in the game.

- **A game situation consisting of two players is called a two person game.**

We will confine our attention to such games.

Games can also be classified according to their outcomes in terms of each player's gains and losses.

- **If the sum of the players' gains and losses is zero then the game is called a zero-sum game.**

In this chapter we will be concerned only with such games. In a **two person zero-sum game** one player's gain is the other player's loss. For example, if one player wins £100 then the other player loses £100, and so the sum of the values is zero.

4.3 Two person zero-sum games

By way of illustration consider the following situation.

A camera company X is planning to introduce a new instant camera and wants to capture as large an increase in its market share as possible. The other dominant camera company is Y. Any gain in market share for X will result in an identical loss in market share for Y. Company Y therefore wishes to minimise X's market share increase. Each company has three plans of action they can follow. The table below shows the percentage increase or decrease in market share for X.

		Y		
		Y_1	Y_2	Y_3
X	X_1	9	7	2
	X_2	11	8	4
	X_3	4	1	−7

This is called the **payoff matrix** or **table**. The payoff table is organised so that the player who is trying to **maximise** the outcome of the game (the offensive player) is down the left-hand side and the player who is trying to minimise the outcome (the defensive player) is across the top. By convention, payoffs are shown for only one of the game players, player X in this case.

In game theory it is assumed that the payoff table is known to both the offensive and defensive players.

A **strategy** is a plan of action to be followed by each player. Each player in a game has two or more strategies, only one of which is selected for each playing of the game. The values in the table are the payoffs or outcomes associated with each player's strategies.

Let us consider two examples.
(i) If X chooses strategy X_2 and Y chooses strategy Y_1 the outcome is an 11% **gain** for X and an 11% **loss** for Y. The number 11 is called the **value** of the game.
(ii) If X chooses strategy X_3 and Y chooses strategy Y_3 the outcome is a **loss** of 7% for X and a 7% **gain** for Y. The **value** of this game is −7.

The purpose of the game is for each player to select the strategy that will result in the best possible payoff or outcome **regardless of what the opponent does**. The best strategy for each player is known as the **optimal strategy**.

When each player in the game adopts a **single strategy** as an optimal strategy then the game is a **pure strategy** game. The value of a pure strategy game is the same for both offensive and defensive players. In contrast, in a **mixed strategy** game the players adopt **different strategies** each time the game is played.

4.4 Pure strategy games

Example 1

Consider the following payoff table in which the row player, X, has three possible strategies, X_1, X_2 and X_3, and the column player, Y, has three possible strategies, Y_1, Y_2 and Y_3. The numbers in the table give points scored when the strategies are adopted.

	Y_1	Y_2	Y_3
X_1	8	12	9
X_2	13	9	8
X_3	11	14	10

In a pure strategy game each player follows a single strategy regardless of the other player's strategy. Each player chooses the strategy that enables him to do the best he can, given that his opponent knows the strategy he is following.

Player X's objective is to obtain as many points as possible.

If player X chooses strategy X_1 (row 1) his minimum points score will be min(8, 12, 9) = 8.

If player X chooses strategy X_2 (row 2) his minimum points score will be min(13, 9, 8) = 8.

If player X chooses strategy X_3 (row 3) his minimum points score will be min(11, 14, 10) = 10.

We may record this by adding an extra column labelled 'row minimum' to the table:

	Y_1	Y_2	Y_3	Row minimum
X_1	8	12	9	8
X_2	13	9	8	8
X_3	11	14	10	10

Since player X wishes to score as many points as possible he will choose the strategy that guarantees the maximum gain, that is player X should choose X_3 since:

$$\text{max}(8, 8, 10) = 10 \tag{1}$$

This is called a **maximin strategy** and
10 = maximum over all rows (row minimum).

To find the **maximin strategy**:
(i)   Find the smallest number in each row. (Remember that -6 is smaller than -1 and -2 is smaller than 1.)
(ii) Select from the numbers found in (i) the largest. The row in which this occurs gives the required strategy.

Player Y's objective is to lose as few points as possible.

If player Y chooses strategy Y_1 (column 1) his maximum loss of points will be max(8, 13, 11) = 13.

If he chooses strategy Y_2 (column 2) his maximum loss of points will be max(12, 9, 14) = 14.

If he chooses strategy Y_3 (column 3) his maximum loss of points will be max(9, 8, 10) = 10.

We may record this by adding an extra row labelled 'column maximum':

	Y_1	Y_2	Y_3
X_1	8	12	9
X_2	13	9	8
X_3	11	14	10
Column maximum	13	14	10

Since player Y wishes to keep his loss of points to a minimum, he will choose the strategy that guarantees this, that is he should choose Y_3 since:

$$\text{min}(13, 14, 10) = 10 \qquad (2)$$

This is called a **minimax strategy** and
10 = minimum over all columns (column maximum).

To find the **minimax strategy**:
(i) Find the largest number in each column. (Remember that 2 is greater than −6 and −1 is greater than −4.)
(ii) Select from the numbers found in (i) the smallest. The column in which this occurs gives the required strategy.

We may include all the above in a single table:

	Y_1	Y_2	Y_3	Row minimum
X_1	8	12	9	8
X_2	13	9	8	8
X_3	11	14	10	10
Column maximum	13	14	10	

$\Big\}$ max = 10

min = 10

Notice that in this case, from equations (1) and (2):

$$\begin{matrix} \text{max over all rows} = & \text{min over all columns} \\ \text{(row minimum)} & \text{(column maximum)} \end{matrix} \qquad (3)$$

- Any two person zero-sum game satisfying equation (3) is said to have a *saddle point*.
- If a game has a saddle point then we call the common value of both sides of equation (3) the *value v* of the game to the row player X.

In this example a saddle point occurs when the row player chooses row 3 and the column player chooses column 3. The numerical value of the saddle point, 10 in this case, is the value of the game. Notice that 10 is **the smallest number in its row** and **the largest number in its column**. Thus, like the centre point of a horse's saddle, a saddle point for a two person zero-sum game is a local minimum in one direction (looking across the row) and a local maximum in another direction (looking up and down the column).

A saddle point can also be thought of as an **equilibrium point** in that neither player can benefit from a unilateral change in strategy. If X, the row player, were to change from his optimal strategy X_3 to either X_1 or X_2 his score would decrease.

If Y, the column player, were to change from his optimal strategy Y_3 to either Y_1 or Y_2 his loss would increase. A saddle point is therefore **stable** in that neither player has an incentive to move away from it.

Many two person zero-sum games do not have a saddle point, as the example below shows. Later in this chapter we will see how to find the value and optimal strategies for two person zero-sum games that do not have saddle points.

Example 2

Show that the two person zero-sum game given by the following payoff table does not have a saddle point.

7	1	4
4	8	11
2	7	9

Calculating the row minimum and the column maximum we obtain:

			Row minimum
7	1	4	1
4	8	11	4
2	7	9	2
Column maximum			
7	8	11	

Maximum of row minimum $= 4$
Minimum of column maximum $= 7$
Since these are not equal there is no saddle point.

Example 3

Obtain the saddle point for the two person zero-sum game given by the following payoff table.

−1	2
−6	−4

The minimum in row 1 is min(−1, 2) = −1.
The minimum in row 2 is min(−6, −4) = −6.
(Note that −6 < −4.)

The maximum of these is −1, since −1 > −6.

The maximum in column 1 is max(−1,−6) = −1.
The maximum in column 2 is max(2,−4) = 2.
The minimum of these is −1.

Hence the saddle point condition is satisfied and (−1) is the saddle point and the value of the game.

Exercise 4A

1 Show that the following two person zero-sum games do not have saddle points.

(a)

3	5	2
4	1	3
1	4	6

(b)

4	5
5	3

(c)

8	6	2	1
7	0	1	3

(d)

6	5
4	8
2	7

2 For the payoff tables below:
   (a) show that there is a saddle point
   (b) obtain the strategies for the players
   (c) give the value of the game.

(i)

5	4
3	2

(ii)

−2	3
−1	1

(iii)

−2	−3
4	−1
1	0

(iv)

5	2	4
−2	1	−1
2	−1	3

(v)

3	6	5	4
4	7	6	5

4.5 Mixed strategies

When there is no saddle point, no pure strategy exists. Players will then play each strategy for a certain proportion (fraction) of the time. This is called a **mixed strategy** game. Let us begin by looking at 2×2 games.

Mixed strategies for 2×2 games

Consider the 2×2 payoff table:

	Y_1	Y_2
X_1	4	2
X_2	3	10

For this game: maximum of row minimum $= 3$
minimum of column maximum $= 4$

so there is **no saddle point**.

Suppose X plays strategy X_1 a fraction $p \, (\geqslant 0)$ of the time, then she will play strategy X_2 a fraction $(1 - p)$ of the time.

The gain for X if Y plays strategy Y_1 is:

$$4p + 3(1 - p) = p + 3$$

The gain for X if Y plays strategy Y_2 is:

$$2p + 10(1 - p) = -8p + 10$$

Let v be the value of the game to X.

That means v is the amount X will win per game if she plays her best strategy consistently and Y plays the best possible counter strategy against her.

Then our problem is to find p so as to maximise v where:

$$v \leqslant p + 3$$
and
$$v \leqslant -8p + 10$$

These inequalities hold since the right-hand sides are what X will win per game against Y's pure strategies Y_1 and Y_2.

Thus we want to find, for each value of p, the smaller of the two functions $(p + 3)$ and $(-8p + 10)$, and then choose the value of p for which the smaller function is as large as possible.

Consider the graphs of the functions $v = p + 3$ and $v = -8p + 10$, for values of p between 0 and 1 (p is a fraction and so $0 \leqslant p \leqslant 1$).

When $p = 0$, $p + 3 = 3$
$p = 1$, $p + 3 = 4$

So $v = p + 3$ goes through $(0, 3)$ and $(1, 4)$.
Similarly, $v = -8p + 10$ goes through $(0, 10)$ and $(1, 2)$.

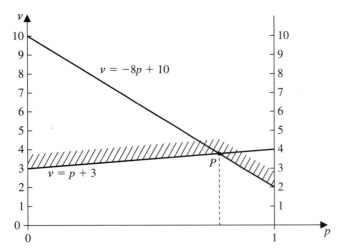

We see that the unshaded region bounded by two line segments, has its highest point at the point P, where the two lines intersect. This must be the case in every such problem.

Hence the required value of p is obtained when:

$$p + 3 = -8p + 10$$

or
$$9p = 7$$

so
$$p = \tfrac{7}{9} \text{ and } (1 - p) = \tfrac{2}{9}$$

X should therefore use strategy X_1 a fraction $\tfrac{7}{9}$ of the time and X_2 a fraction $\tfrac{2}{9}$ of the time. The value v of the game is given by:

$$(p + 3) \quad \text{when} \quad p = \tfrac{7}{9}$$

or
$$-8p + 10 \quad \text{when} \quad p = \tfrac{7}{9}$$

So
$$v = 3\tfrac{7}{9}$$

The point P has coordinates $\left(\tfrac{7}{9}, 3\tfrac{7}{9}\right)$.

We can deal with Y's strategies in the same way. Suppose she uses strategy Y_1 for a fraction q of the time and strategy Y_2 for a fraction $(1 - q)$ of the time.

The gain for Y is then:

$$-[4q + 2(1 - q)] \text{ if X plays strategy } X_1$$

and
$$-[3q + 10(1 - q)] \text{ if X plays strategy } X_2$$

(Notice the minus signs here.)

In the light of the above, q is determined from:

$$4q + 2(1 - q) = 3q + 10(1 - q)$$
$$2q + 2 = -7q + 10$$

or $\qquad 9q = 8$

So $\qquad q = \frac{8}{9}$ and $(1 - q) = \frac{1}{9}$

Y should therefore use strategy Y_1 for $\frac{8}{9}$ of the time and strategy Y_2 for $\frac{1}{9}$ of the time.

Y's loss is given by $-(2q + 2)$ when $q = \frac{8}{9}$ and so is $-3\frac{7}{9}$.

Example 4

The payoff table for a two person zero-sum game is:

	Y_1	Y_2
X_1	2	10
X_2	6	1

Find the best combination of strategies for each player and the value of the game.

Suppose X uses strategy X_1 for a fraction p of the time and strategy X_2 for a fraction $(1 - p)$ of the time. Equating the gain for X when Y plays Y_1, and the gain for X when Y plays Y_2 we obtain:

$$2p + 6(1 - p) = 10p + (1 - p)$$
$$-4p + 6 = 9p + 1$$

or $\qquad 5 = 13p$

So $\qquad p = \frac{5}{13}$

and $\qquad (1 - p) = \frac{8}{13}$

X should therefore use X_1 for a fraction $\frac{5}{13}$ of the time and X_2 for a fraction $\frac{8}{13}$ of the time.
Her gain will be $(9p + 1)$ when $p = \frac{5}{13}$, that is $4\frac{6}{13}$. This is the value of the game.

Suppose Y uses strategy Y_1 for a fraction q of the time and Y_2 for a fraction $(1 - q)$ of the time.

Then:

$$2q + 10(1 - q) = 6q + (1 - q)$$
$$-8q + 10 = 5q + 1$$

or $\qquad 13q = 9$

So $\qquad q = \frac{9}{13}$

and $\qquad (1 - q) = \frac{4}{13}$

The loss of Y will be $-(5q + 1)$ when $q = \frac{9}{13}$, that is $-4\frac{6}{13}$.

Exercise 4B

1 In each of the following 2×2 games find:
   (i)   the optimal mixed strategies for both players
   (ii) the value of the game.

(a)

7	1
4	8

(b)

5	15
11	6

(c)

5	−2
−8	6

Mixed strategies for 2×3 and 3×2 games

Games without saddle points where one player has only two possible strategies can be solved by a straightforward extension of the graphical approach used above.

Example 5

Consider the two person zero-sum game for which the payoff matrix is:

Y

		Y_1	Y_2	Y_3
X	X_1	4	2	6
	X_2	2	10	1

The game does not have a saddle point.

Since X has only two possible strategies suppose he uses strategy X_1 for a fraction p of the time and strategy X_2 for a fraction $(1 - p)$ of the time. Then his expected gain if Y plays Y_1 is:

$$4p + 2(1 - p) = 2 + 2p$$

His expected gain if Y plays Y_2 is:

$$2p + 10(1 - p) = 10 - 8p$$

His expected gain if Y plays Y_3 is:

$$6p + (1 - p) = 1 + 5p$$

So X's problem is to maximise his game value v subject to:

$$v \leqslant 2 + 2p, \; v \leqslant 10 - 8p, \; v \leqslant 1 + 5p$$

by an appropriate choice of p.

We can find the value of p that achieves this by plotting the lines with equations:

$$v = 2 + 2p \; (l_1)$$
$$v = 10 - 8p \; (l_2)$$

and

$$v = 1 + 5p \; (l_3)$$

for

$$0 \leqslant p \leqslant 1$$

These are shown below.

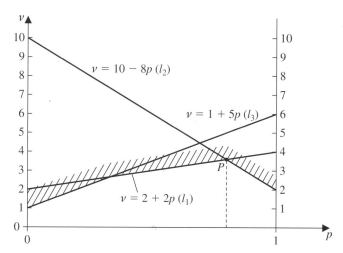

Hence X can maximise the value of the game by choosing the value of p corresponding to the highest point P of the unshaded region. From the diagram, P is at the intersection of l_1 and l_2, so the value of p is given by:

$$2 + 2p = 10 - 8p$$

or

$$10p = 8$$

So

$$p = \tfrac{4}{5} \quad \text{and} \quad (1 - p) = \tfrac{1}{5}$$

and the value of the game is $(2 + 2p)$ when $p = \tfrac{4}{5}$, that is $v = 3\tfrac{3}{5}$.

X should therefore use strategy X_1 for $\tfrac{4}{5}$ of the time and strategy X_2 for $\tfrac{1}{5}$ of the time.

We also note from the above that Y must only use strategies Y_1 and Y_2, corresponding to l_1 and l_2, and never strategy Y_3 if he is to minimise X's gain, when the latter plays the best combination of strategies. This illustrates a general feature of $2 \times n$ games, which is that the player with n strategies will, in fact, be able only to use two of them in his best mix. This follows from the fact that the combination the opponent will use is determined by the intersection of lines corresponding to two strategies. Introducing any strategies other than these two will give the opponent a *greater* gain than necessary.

Having discovered Y will only use strategies Y_1 and Y_2, suppose he uses Y_1 for a fraction q of the time and hence Y_2 for a fraction $(1 - q)$ of the time.

Then using:

(gain for Y if X uses X_1) = (gain for Y if X uses X_2)

we obtain:

$$4q + 2(1 - q) = 2q + 10(1 - q)$$

or
$$2q + 2 = -8q + 10$$
$$10q = 8$$

So
$$q = \tfrac{8}{10} = \tfrac{4}{5} \quad \text{and} \quad (1 - q) = \tfrac{1}{5}$$

Hence Y should use strategy Y_1 for $\tfrac{4}{5}$ of the time and strategy Y_2 for $\tfrac{1}{5}$ of the time and never use Y_3.

Example 6

Consider the two person zero-sum game for which the payoff matrix is:

	Y_1	Y_2
X_1	9	1
X_2	6	5
X_3	3	7

The game does not have a saddle point.

Since Y has only two possible strategies, suppose she uses strategy Y_1 for a fraction q of the time and strategy Y_2 for a fraction $(1 - q)$ of the time. Then her expected gain if X plays X_1 is $-$X's gain:

$$= -[9q + (1 - q)]$$
$$= -[8q + 1]$$

Her expected gain if X plays X_2 is:

$$-[6q + 5(1 - q)]$$
$$= -[q + 5]$$

Her expected gain if X plays X_3 is:

$$-[3q + 7(1 - q)]$$
$$= -[-4q + 7]$$

If v is the value of the game to X, then Y wishes her gain, in each case, to be $\geqslant (-v)$.
So we have

$$-(8q + 1) \geqslant -v$$
or
$$8q + 1 \leqslant v \tag{1}$$

$$-(q + 5) \geqslant -v$$
or
$$q + 5 \leqslant v \tag{2}$$

and
$$-(-4q + 7) \geqslant -v$$
or
$$-4q + 7 \leqslant v \tag{3}$$

Since Y wishes to keep X's gain as small as possible we choose the value of q that satisfies all these constraints and for which v is as small as possible.

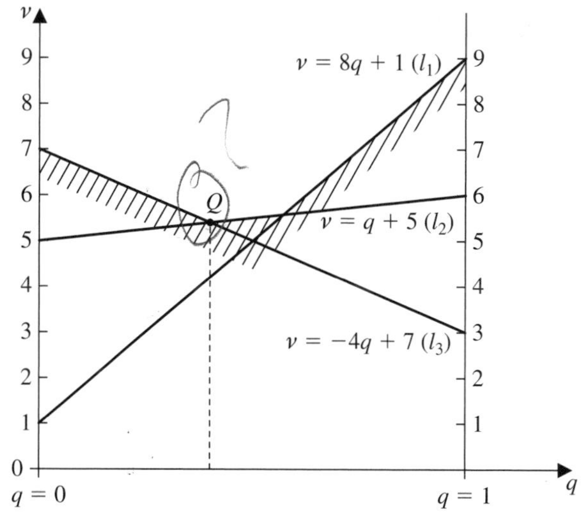

The graph shows the lines:

$$l_1: \quad v = 8q + 1$$
$$l_2: \quad v = q + 5$$
$$l_3: \quad v = -4q + 7$$

The inequalities (1), (2) and (3) are satisfied in the unshaded region and the point Q gives the required value of q. This is the intersection of lines l_2 and l_3.

So $$q + 5 = -4q + 7$$

or $$5q = 2$$

So $$q = \tfrac{2}{5} \quad \text{and} \quad 1 - q = \tfrac{3}{5}$$

Y therefore plays Y_1 for $\tfrac{2}{5}$ of the time and Y_2 for $\tfrac{3}{5}$ of the time. The value of the game is $(q + 5)$ when $q = \tfrac{2}{5}$ and so is $5\tfrac{2}{5}$ to X. This means Y loses $5\tfrac{2}{5}$.

Since the above solution occurs at the intersection of l_2 and l_3, strategy X_1 will not be used by X.

Suppose X uses strategy X_2 for a fraction p of the time and strategy X_3 for a fraction $(1 - p)$ of the time. Then using:

$$(\text{gain for X if Y uses } Y_1) = (\text{gain for X if Y uses } Y_2)$$

$$6p + 3(1 - p) = 5p + 7(1 - p)$$
$$3p + 3 = -2p + 7$$

or $$5p = 4$$

So $$p = \tfrac{4}{5} \quad \text{and} \quad (1 - p) = \tfrac{1}{5}$$

X will therefore not use strategy X_1 at all, will use X_2 a fraction $\tfrac{4}{5}$ of the time and use X_3 a fraction $\tfrac{1}{5}$ of the time.

X's gain will be $(3p + 3)$ when $p = \tfrac{4}{5}$, that is $5\tfrac{2}{5}$, the value of the game to X.

Exercise 4C

1 Find the optimal mixed strategies for both players for each of the games given below.

(a)

5	1	4
1	4	3

(b)

−1	1
1	−2
0	−1

4.6 Formulation of a game as a linear programme

We have seen that it is possible to solve 2×3 and 3×2 games by using graphical methods. However, if we have a 3×3 game then we have to use some other method. In this section we will show that a 3×3 game can be formulated as a linear programme. This can then be solved by using the simplex algorithm discussed in chapter 6 of Book D1. The method we will discuss is a general one, not restricted to 3×3 games, but for any larger problem a computer would almost always be used.

To illustrate the method consider the 3×3 game with payoff matrix:

1	0	−6
−1	−1	2
−2	0	0

This game does not have a saddle point.

We begin by adding a constant to all numbers in the payoff matrix so as to make all entries positive. This makes the value of the new game positive. This ensures that variables in our linear programme are all non-negative, as required by the simplex algorithm. In this case we add 7 to each entry and obtain:

	B_1	B_2	B_3
A_1	8	7	1
A_2	6	6	9
A_3	5	7	7

If the value of the original game was V, then the value of the new game is:

$$v = V + 7$$

For reasons that will become clear in a moment we consider player B first. Suppose he chooses strategies B_1, B_2 and B_3 in the proportions p_1, p_2 and p_3. B wishes to minimise v as this is A's gain.

If A plays strategy A_1, then the expected loss sustained by B is:

$$8p_1 + 7p_2 + p_3$$

Since the value of the game is v, the choice of p_1, p_2 and p_3 must be such as to ensure that **B does not lose more than v** when A plays strategy A_1. Therefore:

$$8p_1 + 7p_2 + p_3 \leqslant v \qquad (1)$$

In a similar way we obtain:

A playing strategy A_2, $\qquad 6p_1 + 6p_2 + 9p_3 \leqslant v \qquad (2)$

A playing strategy A_3, $\qquad 5p_1 + 7p_2 + 7p_3 \leqslant v \qquad (3)$

where $p_1 \geqslant 0$, $p_2 \geqslant 0$ and $p_3 \geqslant 0$.

In addition:

$$p_1 + p_2 + p_3 = 1$$

since p_1, p_2 and p_3 are proportions (or probabilities).

We now divide inequalities (1), (2) and (3) by v (remember that v is positive) and define:

$$x_1 = \frac{p_1}{v}, \quad x_2 = \frac{p_2}{v}, \quad x_3 = \frac{p_3}{v}$$

We then obtain in terms of these new variables:

$$\begin{array}{l} 8x_1 + 7x_2 + x_3 \leqslant 1 \\ 6x_1 + 6x_2 + 9x_3 \leqslant 1 \\ 5x_1 + 7x_2 + 7x_3 \leqslant 1 \end{array} \qquad (4)$$

$$x_1 \geqslant 0, \quad x_2 \geqslant 0, \quad x_3 \geqslant 0$$

and $\qquad\qquad x_1 + x_2 + x_3 = \dfrac{1}{v}$

The objective is to **minimise v**, which means **maximising $\dfrac{1}{v}$**.

The linear programming problem we need to solve is:

maximise $\qquad\qquad P = x_1 + x_2 + x_3$

subject to $\qquad\qquad 8x_1 + 7x_2 + x_3 \leqslant 1$

$\qquad\qquad\qquad\quad 6x_1 + 6x_2 + 9x_3 \leqslant 1$

$\qquad\qquad\qquad\quad 5x_1 + 7x_2 + 7x_3 \leqslant 1$

$\qquad\qquad\quad x_1 \geqslant 0, \quad x_2 \geqslant 0, \quad x_3 \geqslant 0$

This is a maximising problem of the form discussed in chapter 6 of Book D1. It may be solved by introducing slack variables r, s and t so that the inequalities become:

$$8x_1 + 7x_2 + x_3 + r = 1$$
$$6x_1 + 6x_2 + 9x_3 + s = 1$$
$$5x_1 + 7x_2 + 7x_3 + t = 1$$

$$x_1 \geq 0, \quad x_2 \geq 0, \quad x_3 \geq 0, \quad r \geq 0, \quad s \geq 0, \quad t \geq 0$$

The simplex tableaux for solving this problem are shown below.

Initial tableau

Basic variable	x_1	x_2	x_3	r	s	t	Value
r	⑧	7	1	1	0	0	1
s	6	6	9	0	1	0	1
t	5	7	7	0	0	1	1
P	-1	-1	-1	0	0	0	0

Taking ⑧ as the pivot we obtain:

Iteration 1

B variable	x_1	x_2	x_3	r	s	t	Value
x_1	1	$\frac{7}{8}$	$\frac{1}{8}$	$\frac{1}{8}$	0	0	$\frac{1}{8}$
s	0	$\frac{3}{4}$	$\left(8\frac{1}{4}\right)$	$-\frac{3}{4}$	1	0	$\frac{1}{4}$
t	0	$2\frac{5}{8}$	$6\frac{3}{8}$	$-\frac{5}{8}$	0	1	$\frac{3}{8}$
P	0	$-\frac{1}{8}$	$-\frac{7}{8}$	$\frac{1}{8}$	0	0	$\frac{1}{8}$

Taking $\left(8\frac{1}{4}\right)$ as the pivot:

Iteration 2

B variable	x_1	x_2	x_3	r	s	t	Value
x_1	1	$\frac{19}{22}$	0	$\frac{3}{22}$	$-\frac{1}{66}$	0	$\frac{4}{33}$
x_3	0	$\frac{1}{11}$	1	$-\frac{1}{11}$	$\frac{4}{33}$	0	$\frac{1}{33}$
t	0	$\left(\frac{45}{22}\right)$	0	$-\frac{1}{22}$	$-\frac{51}{66}$	1	$\frac{6}{33}$
P	0	$-\frac{1}{22}$	0	$\frac{1}{22}$	$\frac{7}{66}$	0	$\frac{5}{33}$

Taking $\frac{45}{22}$ as the pivot:

Iteration 3

B variable	x_1	x_2	x_3	r	s	t	Value
x_1	1	0	0	$\frac{7}{45}$	$\frac{14}{45}$	$-\frac{19}{45}$	$\frac{2}{45}$
x_3	0	0	1	$-\frac{4}{45}$	$\frac{7}{45}$	$-\frac{2}{45}$	$\frac{1}{45}$
x_2	0	1	0	$-\frac{1}{45}$	$-\frac{17}{45}$	$\frac{22}{45}$	$\frac{4}{45}$
P	0	0	0	$\frac{2}{45}$	$\frac{4}{45}$	$\frac{1}{45}$	$\frac{7}{45}$

This tableau is optimal.

From the final tableau we obtain:

$$x_1 = \tfrac{2}{45}, \quad x_2 = \tfrac{4}{45}, \quad x_3 = \tfrac{1}{45}, \quad P = \tfrac{7}{45}$$

Hence $\dfrac{1}{v} = \frac{7}{45}$, giving $v = \frac{45}{7} = 6\frac{3}{7}$

and:
$$p_1 = x_1 v = \left(\tfrac{2}{45}\right)\left(\tfrac{45}{7}\right) = \tfrac{2}{7}$$
$$p_2 = x_2 v = \left(\tfrac{4}{45}\right)\left(\tfrac{45}{7}\right) = \tfrac{4}{7}$$
$$p_3 = x_3 v = \left(\tfrac{1}{45}\right)\left(\tfrac{45}{7}\right) = \tfrac{1}{7}$$

The value of the original game $V = v - 7 = 6\frac{3}{7} - 7 = -\frac{4}{7}$

The optimal policy for B is:

play strategy B_1 for $\frac{2}{7}$ of the time

play strategy B_2 for $\frac{4}{7}$ of the time

play strategy B_3 for $\frac{1}{7}$ of the time

Let us now consider the game from A's point of view. Suppose A chooses strategies A_1, A_2 and A_3 in the proportions q_1, q_2 and q_3. The value v of the game to A cannot be greater than his gain under each of B's strategies.
Hence we have:

B playing strategy B_1, $8q_1 + 6q_2 + 5q_3 \geqslant v$
B playing strategy B_2, $7q_1 + 6q_2 + 7q_3 \geqslant v$
B playing strategy B_3, $q_1 + 9q_2 + 7q_3 \geqslant v$

where:
$$q_1 + q_2 + q_3 = 1,$$
$$q_1 \geqslant 0, \quad q_2 \geqslant 0, \quad q_3 \geqslant 0$$

and v is to be maximised.
Dividing by v and defining:

$$y_1 = \frac{q_1}{v}, \quad y_2 = \frac{q_2}{v}, \quad y_3 = \frac{q_3}{v}$$

we obtain:
$$8y_1 + 6y_2 + 5y_3 \geqslant 1$$
$$7y_1 + 6y_2 + 7y_3 \geqslant 1$$
$$y_1 + 9y_2 + 7y_3 \geqslant 1$$

where:
$$y_1 + y_2 + y_3 = \frac{1}{v}$$

$$y_1 \geqslant 0, \quad y_2 \geqslant 0, \quad y_3 \geqslant 0$$

The objective is to **maximise** v, which means **minimising** $\frac{1}{v}$.

The linear programming problem to be solved to obtain A's strategy is then:

minimise
$$Q = y_1 + y_2 + y_3$$

subject to
$$8y_1 + 6y_2 + 5y_3 \geqslant 1$$
$$7y_1 + 6y_2 + 7y_3 \geqslant 1 \tag{5}$$
$$y_1 + 9y_2 + 7y_3 \geqslant 1$$

$$y_1 \geqslant 0, \quad y_2 \geqslant 0, \quad y_3 \geqslant 0$$

This is a minimising problem with $\geqslant$ inequalities. This kind of problem was not dealt with in Book D1 as it requires a much more elaborate method for its solution. However, two things should be noted.

(i) The matrix of coefficients on the left-hand side of inequalities (4) is:

$$\begin{pmatrix} 8 & 7 & 1 \\ 6 & 6 & 9 \\ 5 & 7 & 7 \end{pmatrix} \tag{6}$$

The matrix of coefficients on the left-hand side of inequalities (5) is:

$$\begin{pmatrix} 8 & 6 & 5 \\ 7 & 6 & 7 \\ 1 & 9 & 7 \end{pmatrix} \tag{7}$$

This is the transpose of (6) above. In fact the linear programming problem to be solved to find A's strategy is the **dual** of the linear programming problem to be solved to find B's strategy. The dual of a linear programming problem is defined in Appendix 1.

(ii) **We can in fact obtain the strategy A should use from the final tableau of the calculation for B's strategy.**
The final row of that tableau was:

	x_1	x_2	x_3	r	s	t	Value
P	0	0	0	$\frac{2}{45}$	$\frac{4}{45}$	$\frac{1}{45}$	$\frac{7}{45}$

The entry under r gives the value of y_1.
The entry under s gives the value of y_2.
The entry under t gives the value of y_3.

So
$$y_1 = \tfrac{2}{45}, \quad y_2 = \tfrac{4}{45}, \quad y_3 = \tfrac{1}{45}$$

and $Q = \dfrac{1}{v} = \tfrac{7}{45} \Rightarrow v = \tfrac{45}{7}$, as before.

$$\therefore \qquad q_1 = vy_1 = \tfrac{45}{7}\left(\tfrac{2}{45}\right) = \tfrac{2}{7}$$
$$q_2 = vy_2 = \tfrac{45}{7}\left(\tfrac{4}{45}\right) = \tfrac{4}{7}$$
$$q_3 = vy_3 = \tfrac{45}{7}\left(\tfrac{1}{45}\right) = \tfrac{1}{7}$$

Hence A should use strategy A_1 for $\tfrac{2}{7}$ of the time, strategy A_2 for $\tfrac{4}{7}$ of the time and strategy A_3 for $\tfrac{1}{7}$ of the time. As before, since $v = 6\tfrac{3}{7}$, the value V of the original game is $6\tfrac{3}{7} - 7 = -\tfrac{4}{7}$.

Example 7

		B		
		I	II	III
A	I	4	2	6
	II	2	10	1

The payoff matrix for a two person zero-sum game is shown above.

(a) Obtain the linear programming problem which when solved will give the optimal strategy for B.
(b) Solve this linear programming problem to obtain B's optimal strategy and the value of the game.
(c) Deduce the optimal strategy for A.

(a) Suppose B chooses strategies $I(B_1)$, $II(B_2)$ and $III(B_3)$ in proportions p_1, p_2 and p_3 then:

$$4p_1 + 2p_2 + 6p_3 \leqslant v$$
$$2p_1 + 10p_2 + p_3 \leqslant v$$

where v is the value of game and

$$p_1 + p_2 + p_3 = 1$$

Define $x_1 = \dfrac{p_1}{v}$, $x_2 = \dfrac{p_2}{v}$, $x_3 = \dfrac{p_3}{v}$. Then the linear programming problem is:

maximise
$$P = x_1 + x_2 + x_3$$

subject to
$$4x_1 + 2x_2 + 6x_3 \leqslant 1$$
$$2x_1 + 10x_2 + x_3 \leqslant 1$$
$$x_1 \geqslant 0, \quad x_2 \geqslant 0, \quad x_3 \geqslant 0$$

(b) Initial tableau:

B variable	x_1	x_2	x_3	r	s	Value
r	④	2	6	1	0	1
s	2	10	1	0	1	1
P	-1	-1	-1	0	0	0

Using ④ as the pivot the next tableau is:

B variable	x_1	x_2	x_3	r	s	Value
x_1	1	$\frac{1}{2}$	$1\frac{1}{2}$	$\frac{1}{4}$	0	$\frac{1}{4}$
s	0	⑨	-2	$-\frac{1}{2}$	1	$\frac{1}{2}$
P	0	$-\frac{1}{2}$	$\frac{1}{2}$	$\frac{1}{4}$	0	$\frac{1}{4}$

Using ⑨ as the pivot the next tableau is:

B variable	x_1	x_2	x_3	r	s	Value
x_1	1	0	$\frac{29}{18}$	$\frac{5}{18}$	$-\frac{1}{18}$	$\frac{2}{9}$
x_2	0	1	$-\frac{2}{9}$	$-\frac{1}{18}$	$\frac{1}{9}$	$\frac{1}{18}$
P	0	0	$\frac{7}{18}$	$\frac{2}{9}$	$\frac{1}{18}$	$\frac{5}{18}$

This tableau is optimal.
From the final tableau:

$$x_1 = \tfrac{2}{9}, \quad x_2 = \tfrac{1}{18}, \quad x_3 = 0, \quad P = \frac{1}{v} = \tfrac{5}{18}$$

Hence
$$v = \tfrac{18}{5} = 3\tfrac{3}{5}, \text{ the value of the game,}$$

and
$$p_1 = vx_1 = \tfrac{18}{5}\left(\tfrac{2}{9}\right) = \tfrac{4}{5}$$

$$p_2 = vx_2 = \tfrac{18}{5}\left(\tfrac{1}{18}\right) = \tfrac{1}{5}$$

$$p_3 = 0$$

Therefore the optimal policy for B is:

play strategy I(B_1) for $\frac{4}{5}$ of the time

play strategy II(B_2) for $\frac{1}{5}$ of the time

never play strategy III(B_3).

Also, from the last line of the final tableau we can see, using the same notation as above, that:

$$y_1 = \tfrac{2}{9}, \quad y_2 = \tfrac{1}{18}$$

and
$$Q = \frac{1}{v} = \tfrac{5}{18}$$

so
$$v = \tfrac{18}{5} = 3\tfrac{3}{5}$$

Therefore
$$q_1 = vy_1 = \tfrac{18}{5}\left(\tfrac{2}{9}\right) = \tfrac{4}{5}$$

and
$$q_2 = vy_2 = \tfrac{18}{5}\left(\tfrac{1}{18}\right) = \tfrac{1}{5}$$

The optimal policy for A is therefore:

play strategy I(A_1) for $\tfrac{4}{5}$ of the time

play strategy II(A_2) for $\tfrac{1}{5}$ of the time

The value of the game is $3\tfrac{3}{5}$.

Note: Two alternative methods for formulating a game as a linear programming problem are given in Appendix 2.

Exercise 4D

1

	B		
	I	**II**	**III**
I	4	−1	−3
II	−2	0	3

A

For the payoff matrix given above:
(a) obtain the linear programming problem which when solved will give the optimal strategy for B
(b) solve the linear programming problem to obtain B's optimal strategy and the value of the game
(c) deduce the optimal strategy for A.

2 The payoff matrix for a 3×3 game is:

	B		
	I	**II**	**III**
I	8	4	2
II	2	8	4
III	2	1	8

A

(a) Obtain the linear programming problem which when solved will give the optimal strategy for B.

(b) Solve the linear programming problem to obtain B's optimal strategy and the value of the game.

(c) Deduce the optimal strategy for A.

4.7 Dominance

The principle of **dominance** can be used to reduce the size of games by eliminating strategies that would never be used. A strategy for a player can be eliminated if the player can always do as well or better playing another strategy. In other words, **a strategy can be eliminated if all its game outcomes are the same or worse than the corresponding game outcomes of another strategy.**

Example 8

Let us look again at the payoff table given in Example 1.

Y

		Y_1	Y_2	Y_3
	X_1	8	12	9
X	X_2	13	9	8
	X_3	11	14	10

For player X's strategies we employ **the dominance rule for rows:**

■ **Every value in the dominating row must be equal to, or *greater than*, the corresponding value in the dominated row.**

Applying this rule to the above table we can see that row 3, that is strategy X_3, dominates row 1, that is strategy X_1, since $11 > 8$, $14 > 12$ and $10 > 9$.

Row 1 and strategy X_1 can therefore be eliminated. The reduced payoff matrix obtained is:

Y

		Y_1	Y_2	Y_3
X	X_2	13	9	8
	X_3	11	14	10

For player Y's strategies we employ **the dominance rule for columns:**

■ **Every value in the dominating column must be equal to, or *less than*, the corresponding value in the dominated column.**

Note: The rule for rows involves '**greater than**' but the rule for columns involves '**less than**'.

Applying the column rule to the reduced payoff table above we see that column 3, that is strategy Y_3, **dominates** column 2, that is strategy Y_2, since $8 < 9$ and $10 < 14$.

Column 2 can therefore be eliminated. The reduced payoff matrix is then:

	Y_1	Y_3
X_2	13	8
X_3	11	10

We have reduced the original 3×3 game to a 2×2 game. Clearly it is easier to handle the smaller game. This game has a saddle point since:

$$\text{max(row minimum)} = 10$$
$$\text{min(column maximum)} = 10$$

Example 9

Consider the following payoff table:

		B		
		I	II	III
	I	8	6	2
A	II	11	8	4
	III	4	1	7

(a) Apply the principle of dominance to this payoff table.
(b) Determine player A's optimal strategy.
(c) Determine player B's optimal strategy.
(d) State the value of the game.

(a) Consider first the rows of the table. Row 2 dominates row 1 since $11 > 8$, $8 > 6$ and $4 > 2$.
So we can eliminate row 1.
The reduced table is then:

		B		
		I	II	III
A	II	11	8	4
	III	4	1	7

Consider now the columns.
Column 2 dominates column 1 since $8 < 11$ and $1 < 4$.
So we can eliminate column 1 and obtain the reduced table:

B

		II	III
A	II	8	4
	III	1	7

(b) Suppose A uses strategy II for a fraction p of the time and strategy III for a fraction $(1 - p)$ of the time. Then we require that:

$$8p + (1 - p) = 4p + 7(1 - p)$$

so

$$7p + 1 = -3p + 7$$

and

$$10p = 6$$

or

$$p = \tfrac{3}{5} \text{ and } (1 - p) = \tfrac{2}{5}$$

The optimal strategy for A is then:

> never use strategy I
>
> use strategy II for $\tfrac{3}{5}$ of the time
>
> use strategy III for $\tfrac{2}{5}$ of the time

(c) Suppose B uses strategy II for a fraction q of the time and strategy III for a fraction $(1 - q)$ of the time. Then we require that:

$$8q + 4(1 - q) = q + 7(1 - q)$$

so

$$4q + 4 = -6q + 7$$

and

$$10q = 3$$

or

$$q = \tfrac{3}{10} \text{ and } (1 - q) = \tfrac{7}{10}$$

The optimal strategy for B is then:

> never use strategy I
>
> use strategy II for $\tfrac{3}{10}$ of the time
>
> use strategy III for $\tfrac{7}{10}$ of the time

(d) The value of the game is given, from (b), by $(7p + 1)$ when $p = \tfrac{3}{5}$, so $v = \tfrac{26}{5} = 5\tfrac{1}{5}$.

We could also calculate it from (c) as $(4q + 4)$ when $q = \tfrac{3}{10}$, which again gives $v = 5\tfrac{1}{5}$.

4.8 Summary of how to solve a two person zero-sum game

■ To end this chapter we present a summary of the procedure for finding the optimal strategies and the value of a two person zero-sum game.

Step 1 Check to see if the game has a saddle point. If not go to step 2.

Step 2 Eliminate as many rows and columns as possible using the principle of dominance.

Step 3 Solve the resulting game by an appropriate method. If we have a 2×2, 2×3 or 3×2 game these may be solved graphically as we have seen. Otherwise solve the game by using linear programming methods.

Exercise 4E

1 Consider the following payoff table:

B

		I	II	III
	I	1.8	1.1	1.3
A	II	1.3	1.5	1.8
	III	1.3	1.3	1.8

(a) Apply the principle of dominance to this payoff table.

(b) Determine player A's optimal strategy.

(c) Determine player B's optimal strategy.

(d) State the value of the game.

2 Consider the following payoff table:

Y

		I	II	III
	I	−2	4	5
X	II	0	−3	6
	III	−5	1	−6

(a) Apply the principle of dominance to this payoff table.

(b) Determine player X's optimal strategy.

(c) Determine player Y's optimal strategy.

(d) State the value of the game.

3 Consider the following payoff matrix:

Q

	I	II
I	2	4
II	1	2
III	3	0
IV	−1	6

P

(a) Apply the principle of dominance to this payoff table.

(b) Determine player Q's optimal strategy.

(c) Hence determine player P's optimal strategy.

(d) What is the value of the game?

SUMMARY OF KEY POINTS

1 A **two person game** is one in which only two parties can play.

2 A **zero-sum game** is one in which the sum of the losses for one player is equal to the sum of the gains for the other player.

3 A two person zero-sum game for which max over all rows (row minimum) = min over all columns (column maximum) is said to have a **saddle point**. The common value of both sides of the above equation is the **value** v of the game to the row player.

4 **Dominance rule for rows**
If every entry in a given row R_1 is **less than or equal to** the corresponding entry in another given row R_2, then R_1 is said to be **dominated** by R_2 and may therefore be eliminated.

5 **Dominance rule for columns**
If every entry in a given column C_1 is **greater than or equal to** the corresponding entry in another given column C_2, then C_1 is said to be **dominated** by C_2 and may therefore be eliminated.

6 To solve a two person zero-sum game.
Step 1   Check to see if the game has a saddle point. If not go to step 2.
Step 2   Eliminate as many rows and columns as possible using the principle of dominance.
Step 3   Solve the resulting game by an appropriate method.

Dynamic programming

5

5.1 What is dynamic programming?

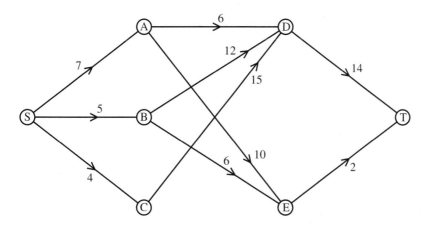

The diagram above shows a directed network. In this network, as in many other networks used to model real-world problems, it is possible to identify **distinct stages**. At each stage a decision has to be made regarding in which direction to go. At S we are faced with three possible routes: go to A or to B or to C. Having made a choice we move to the next stage. Suppose we are at B. Then we are presented with two possible choices. Which choice is made is determined by the overall objective. Let us for the moment take as our objective 'to find a route from S to T so that the total length is a minimum.'

Dynamic programming (D.P.) is a technique for solving **multi-stage decision-making problems** of this kind. It enables **optimal solutions** to be obtained. The technique was developed by Richard Bellman in the 1950s and since then has been applied in many areas, including stock control, allocation of resources, routing problems and equipment replacement and maintenance.

5.2 The principle of optimality

The problem of finding the shortest route from S to T in the above network when the weights are distances was considered in chapter 3 of Book D1. There it was solved by using Dijkstra's algorithm. If you apply that algorithm to this problem you obtain SBET as the shortest route of length 13. In addition, you can see that:

the shortest route from S to E is SBE   (11)

the shortest route from S to B is SB   (5)

Any part of the shortest route from S to T is itself a shortest path. This is an example of the **principle of optimality** first postulated by Bellman:

- **Any part of an optimal path is itself optimal.**

We will see later in this chapter that, depending on the problem under consideration, 'optimal' has different meanings.

This principle is the basis of the dynamic programming technique.

5.3 Dynamic programming applied to the shortest route problem

For ease of reference we repeat here the network drawn in section 5.1.

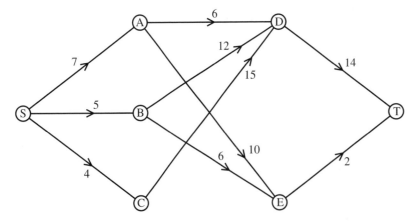

Whereas the Dijkstra algorithm began at S, the dynamic programming technique works backwards from T to S.

We begin by considering the vertices joined directly to T, namely D and E – these are **stage 1** vertices. The best route from these to T is noted. We now move to the next set of vertices that are joined directly to D and E, namely vertices A, B and C – these are **stage 2** vertices. The best route from these to T is found using the optimal routes from the stage 1 vertices. This process is repeated once again when S is reached. The principle of optimality is used at each stage and the current optimal path is obtained by using the previously obtained optimal paths.

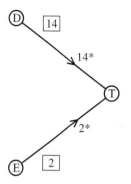

■ **Stage 1**

From D there is no choice and the distance DT is 14. Similarly from E there is no choice and the distance ET $= 2$.

We therefore label D with $\boxed{14}$ and E with $\boxed{2}$ as these are the lengths of the shortest routes to T. Also, as both DT and ET are optimal, we indicate this with a star.

■ **Stage 2**

(i) From A to T there are two possible routes, ADT and AET:

length ADT $=$ length AD $+$ label of D $= 6 + 14 = 20$

length AET $=$ length AE $+$ label of E $= 10 + 2 = 12$

Since we are looking for the **shortest route**, A is assigned the label $\boxed{12} = \min(20, 12)$.

(ii) From B to T there are two possible routes, BDT and BET:

length BDT $=$ length BD $+$ label of D $= 12 + 14 = 26$

length BET $=$ length BE $+$ label of E $= 6 + 2 = 8$

Since we are looking for the **shortest route**, B is assigned the label $\boxed{8} = \min(26, 8)$.

(iii) From C to T there is only one route, CDT:

length CDT $=$ length CD $+$ label of D $= 15 + 14 = 29$

So C is assigned the label $\boxed{29}$.

We then have:

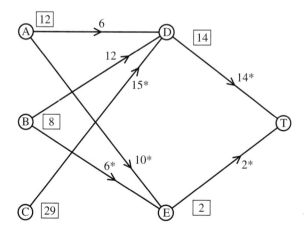

The stars indicate the optimal routes.

■ **Stage 3**

From S there are three possible choices: we may choose a route through A or B or C.

(i) If we choose A, the shortest route has length:

$$\text{length SA} + \text{label of A} = 7 + 12 = 19$$

(ii) If we choose B, the shortest route through B has length:

$$\text{length SB} + \text{label of B} = 5 + 8 = 13$$

(iii) If we choose C, the shortest route through C has length:

$$\text{length SC} + \text{label of C} = 4 + 29 = 33$$

The shortest route then passes through B and is of length $13 = \min(19, 13, 33)$.

We then have:

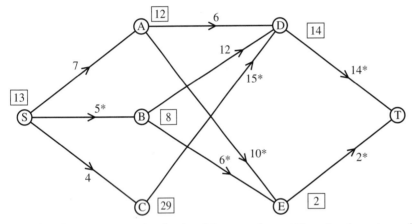

The shortest route is obtained by starting at S and using starred edges. It is therefore SB, BE and ET:

$$S \rightarrow B \rightarrow E \rightarrow T$$

5.4 Some terminology used in dynamic programming

The **stage** tells us how far the vertex in question is from the destination vertex T. In our example:

> D and E are **stage 1** vertices
> A, B and C are **stage 2** vertices
> S is a **stage 3** vertex

Within each stage there are various possible **states**. For example, A, B and C are states within stage 2.

A stage–state notation $(\alpha; \beta)$ is sometimes used where α is the stage and β is the state. Using this notation:

$$T \equiv (0;\ 1),\ D \equiv (1;\ 1),\ E \equiv (1;\ 2)$$
$$A \equiv (2;\ 1),\ B \equiv (2;\ 2),\ C \equiv (2;\ 3),\ S \equiv (3;\ 1)$$

The possible choices at each vertex are called **actions**. For example, at B the possible actions are BD and BE. These are directed arcs from one state to the next.

Values are the numbers calculated for each state at each stage. In general these will depend on the objective. For the above example the optimal value is the smallest value. The optimal value is the **label** assigned to the vertex.

Using this terminology the above network becomes:

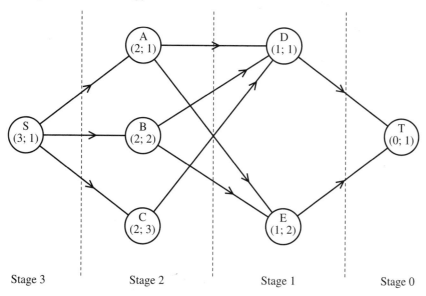

Stage 3   Stage 2   Stage 1   Stage 0

The calculations described above may also conveniently be put in concise tabular form:

Stage	Initial state	Action	Destination	Value
1	D	DT	T	14*
	E	ET	T	2*
2	A	AD	D	$6 + 14 = 20$
		AE	E	$10 + 2 = 12$*
	B	BD	D	$12 + 14 = 26$
		BE	E	$6 + 2 = 8$*
	C	CD	D	$15 + 14 = 29$*
3	S	SA	A	$7 + 12 = 19$
		SB	B	$5 + 8 = 13$*
		SC	C	$4 + 29 = 33$

Here * indicates the optimal value (smallest).

Hence the shortest route is of length 13. Reading *up* the table, the edges on the shortest route are SB, BE and ET, and so the route is $S \rightarrow B \rightarrow E \rightarrow T$, (SBET).

In the examination you will often be given a table of this kind to complete. If you are not given one you should seriously consider drawing up such a table as it makes the recording of your calculations very straightforward and also enables you to obtain the optimal route very quickly and easily.

5.5 Dynamic programming applied to the longest route problem

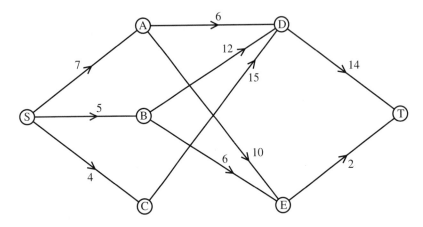

The above network considered earlier may also be a model of the possible routes that a salesman can take from S to T via various cities. The weights, however, now give the commission, in units of £10, that he can earn on particular sections of his journey. The objective then is to choose the route that **maximises** the commission. We are therefore looking for the **longest route through the network**.

We will present the calculations in tabular form. The first four columns of the table are as in the shortest route calculation since the **stages** are the same, the **states** are the same, the **actions** (possible choices at each vertex) are the same and the **destinations** are the same. In completing the value column we choose the **largest** value, indicated by *, since we are now looking for the **longest** route.

Stage	Initial state	Action	Destination	Value
1	D	DT	T	14*
	E	ET	T	2*
2	A	AD	D	$6 + 14 = 20*$
		AE	E	$10 + 2 = 12$
	B	BD	D	$12 + 14 = 26*$
		BE	E	$6 + 2 = 8$
	C	CD	D	$15 + 14 = 29*$
3	S	SA	A	$7 + 20 = 27$
		SB	B	$5 + 26 = 31$
		SC	C	$4 + 29 = 33*$

The longest route is of length 33. Reading up the table the edges on the longest route are SC, CD and DT, and so the route is $S \rightarrow C \rightarrow D \rightarrow T$, (SCDT). The maximum commission is obtained by using this route and is £330.

Exercise 5A

1 Use dynamic programming to find:
   (i) the shortest route
   (ii) the longest route
       from S to T in each of the following networks.
   (a)

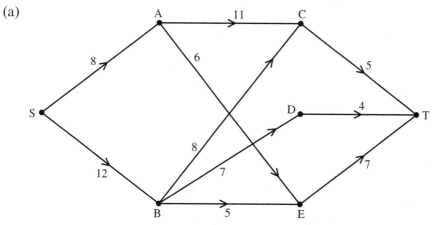

(b)

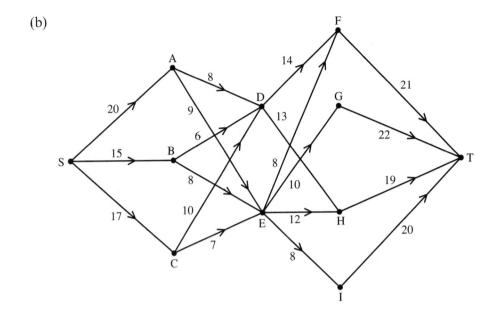

5.6 Some other routing problems

There are two other routing problems that we can solve using dynamic programming. These arise for different objective criteria.

The minimax route

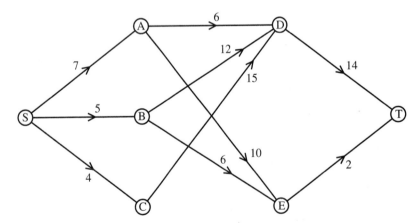

We will again consider the same network as before. Now, however, the weight on an edge gives the maximum altitude on that road, in 1000 ft, above sea level. A traveller wishing to travel from S to T wants to choose a route so as to minimise the maximum altitude above sea level along the route. Such a route is called a **minimax route**. In general:

■ **A minimax route is one on which the maximum length of the edges used is as small as possible.**

We again present the calculations in tabular form. The first four columns are as in the previous examples. The value column requires some explanation. The values are now lengths of edges.

Stage	Initial state	Action	Destination	Value
1	D	DT	T	14*
	E	ET	T	2*
2	A	AD	D	max(6, 14) = 14
		AE	E	max(10, 2) = 10*
	B	BD	D	max(12, 14) = 14
		BE	E	max(6, 2) = 6*
	C	CD	D	max(15, 14) = 15*
3	S	SA	A	max(7, 10) = 10
		SB	B	max(5, 6) = 6*
		SC	C	max(4, 15) = 15

Let us look at the various stages in turn.

Stage 1

Stage	Initial state	Action	Destination	Value
1	D	DT	T	14*
	E	ET	T	2*

As in previous cases, since DT is the only edge from D to T its length, 14, is both maximum and minimum. Similarly, the length 2 of ET is both maximum and minimum.
Both of these are therefore indicated by a star.

Stage 2

Stage	Initial state	Action	Destination	Value
2	A	AD	D	max(6, 14) = 14
		AE	E	max(10, 2) = 10*
	B	BD	D	max(12, 14) = 14
		BE	E	max(6, 2) = 6*
	C	CD	D	max(15, 14) = 15*

The first line refers to path ADT. On this path there are two edges: AD, length 6, and DT, length 14. The maximum length of an edge on this path is then:

$$\max(6, 14) = 14 \tag{1}$$

The second line refers to path AET. On this path there are two edges: AE, length 10, and ET, length 2. The maximum length of an edge on this path is then:

$$\max(10, 2) = 10 \qquad (2)$$

The minimum of equations (1) and (2) is 10. This is the minimum value of the maximum edge lengths on all possible routes from A to T. We have therefore starred the 10. The calculations for initial states B and C are carried out in a similar manner.

Stage 3

Stage	Initial state	Action	Destination	Value
3	S	SA	A	$\max(7, 10) = 10$
		SB	B	$\max(5, 6) = 6^*$
		SC	C	$\max(4, 15) = 15$

(i) The first line refers to routes from S to T via A. Using the starred value 10 for initial state A, obtained in stage 2, we see that the minimum value of maximum edge lengths on all such routes is:

$$\max(7, 10) = 10 \qquad (3)$$

since length SA = 7.

(ii) The second line refers to routes from S to T via B. Using the starred value 6 for initial state B, obtained in stage 2, we see that the minimum value of maximum edge lengths on all such routes is:

$$\max(5, 6) = 6 \qquad (4)$$

since length SB = 5.

(iii) The third line refers to routes from S to T via C. Using the starred value 15 for initial state C, obtained in stage 2, we see that the minimum value of maximum edge lengths on all such routes is:

$$\max(4, 15) = 15 \qquad (5)$$

The minimum of the values found in equations (3), (4) and (5) is the 6 found in equation (4). The route we are looking for is therefore such that the maximum length of edge used is 6 – this is the edge BE.

Working upwards through the table from stage 3 to stage 1 we can obtain this route. From stage 3 the action is SB. From stage 2, with initial state B, the action is BE. From stage 1, with initial state E, the action is ET. The **minimax route** is therefore SBET. On this route the maximum edge length 6 (BE = 6) is a minimum.

The route the traveller should use is SBET and on this route the maximum altitude encountered will be 6000 ft above sea level.

The maximin route

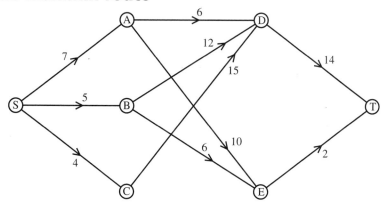

We now take the above network to model possible production lines, starting at S and finishing at T. The weights are now production rates and the objective is to determine the route for which the slowest leg is as fast as possible.

Such a route is called a **maximin route**. In general:

■ **A maximin route is one for which the minimum length of the edges used is as large as possible.**

We again present the calculations in tabular form. The first four columns are as in the previous examples. The numbers in the value column again refer to edge lengths. In the calculations the roles of maximum and minimum are reversed. Otherwise the calculation proceeds exactly as in the minimax problem.

Stage	Initial state	Action	Destination	Value
1	D	DT	T	14*
	E	ET	T	2*
2	A	AD	D	min(6, 14) = 6*
		AE	E	min(10, 2) = 2
	B	BD	D	min(12, 14) = 12*
		BE	E	min(6, 2) = 2
	C	CD	D	min(15, 14) = 14*
3	S	SA	A	min(7, 6) = 6*
		SB	B	min(5, 12) = 5
		SC	C	min(4, 14) = 4

Working upwards through the table from stage 3 to stage 1 we see that the route uses:

> stage 3: SA
> stage 2: AD
> stage 1: DT

On this route the **minimum** edge length (AD = 6) is a **maximum**. The route SADT is the one that should be used, then the minimum rate will be 6.

Exercise 5B

1

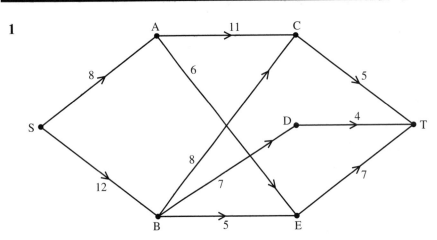

For the above network find:

(a) the minimax route from S to T

(b) the maximin route from S to T.

5.7 Other applications of dynamic programming

In the three examples that follow we show how dynamic programming may be applied to problems, other than routing problems, where distinct stages can be identified. We also illustrate how to tackle problems when the information is not given as a network.

Example 1

At the beginning of each month an advertising manager must choose one of three adverts: the previous advert, the current advert or a new advert.

She therefore has three options:

        A   use the previous advert
        B   use the current advert
        C   run a new advert.

The possible choices are shown in the network below, together with the expected profits, in thousands of pounds, on the edges.

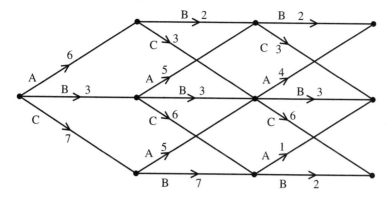

The manager wants to maximise her profits for the 3-month period. Obtain the sequence of decisions she should make in order to obtain the maximum profit.

In this case we begin by labelling the vertices with (stage; state) variables:

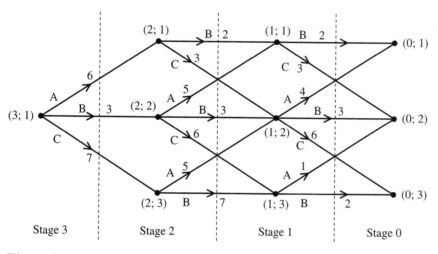

The calculations are conveniently carried out in a table. An additional column has been added to record total profit.

Stage	State	Action	Profit	Total profit
1	1	B	2	2
		C	3	3*
	2	A	4	4
		B	3	3
		C	6	6*
	3	A	1	1
		B	2	2*

Stage	State	Action	Profit	Total profit
2	1	B to (1; 1)	2	$2 + 3 = 5$
		C to (1; 2)	3	$3 + 6 = 9*$
	2	A to (1; 1)	5	$5 + 3 = 8$
		B to (1; 2)	3	$3 + 6 = 9*$
		C to (1; 3)	6	$6 + 2 = 8$
	3	A to (1; 2)	5	$5 + 6 = 11*$
		B to (1; 3)	7	$7 + 2 = 9$
3	1	A to (2; 1)	6	$6 + 9 = 15$
		B to (2; 2)	3	$3 + 9 = 12$
		C to (2; 3)	7	$7 + 11 = 18*$

* indicates maximum

The sequence of decisions required is obtained by reading upwards through the table:

stage 3   C to (2; 3)
stage 2   A to (1; 2)
stage 1   C to (0; 3)

Profit is £18 000.

Example 2

A builder is planning to build three new buildings, A, B and C, at the rate of one per year. The order in which they are to be built is a matter of choice. The estimates of costs made by the builder are shown in the table below, in units of £1000.

Cost (in units of £1000)

Already built	A	B	C
Nothing	60	50	40
A	—	55	45
B	70	—	50
C	65	60	—
A and B	—	—	55
A and C	—	65	—
B and C	75	—	—

For tax reasons, it is advantageous to arrange the sequence of building so that the least annual cost is as large as possible. Determine the order in which the builder should build the buildings.

As a first step we draw a network to model the situation and at the same time introduce (stage; state) variables.

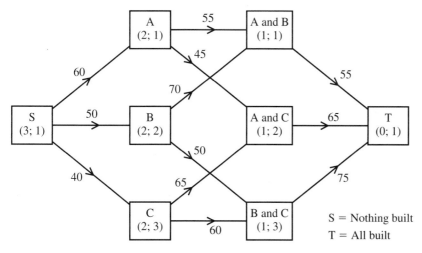

The route from S to T that we require in this network is the maximin route. The calculations are shown in the table below.

Stage	Initial state	Action	Destination	Value
1	(1; 1)	Build C	T(0; 1)	55*
	(1; 2)	Build B	T(0; 1)	65*
	(1; 3)	Build A	T(0; 1)	75*
2	(2; 1)	Build B	(1; 1)	min(55, 55) = 55*
	A	Build C	(1; 2)	min(45, 65) = 45
	(2; 2)	Build A	(1; 1)	min(70, 55) = 55*
	B	Build C	(1; 3)	min(50, 75) = 50
	(2; 3)	Build A	(1; 2)	min(65, 65) = 65*
	C	Build B	(1; 3)	min(60, 75) = 60
3	(3; 1)	Build A	(2; 1)	min(60, 55) =55*
	S	Build B	(2; 2)	min(50, 55) = 50
		Build C	(2; 3)	min(40, 65) = 40

*indicates maximum

Reading upwards through the table:

<pre>
stage 3   build A   [go to (2; 1)]
stage 2   build B   [go to (1; 1)]
stage 1   build C   [go to (0; 1)]
</pre>

The buildings should therefore be built in the order A then B then C. On this route the least annual cost is 55.

Example 3

A salesman has a list of eight stores on his list. He plans to visit only one store per day for 3 days and so divides his stores into groups, depending on the parent company. His plan is:

Monday	Tuesday	Wednesday
Group 1	Group 2	Group 3
A, B, C	D, E	F, G, H

The costs of travelling between home and the various stores are shown in the table below.

	Home (T)	**A**	**B**	**C**	**D**	**E**	**F**	**G**	**H**
Home (S)		20	19	17					
A					22	14			
B					26	28			
C					24	30			
D							18	19	20
E							22	16	18
F	42								
G	40								
H	45								

The salesman wishes to arrange his visits so that, starting and finishing at home, his total costs are a minimum. Determine the route he should take and give its cost.

As a first step we draw a network modelling the situation with home both the start vertex S and the destination vertex T.

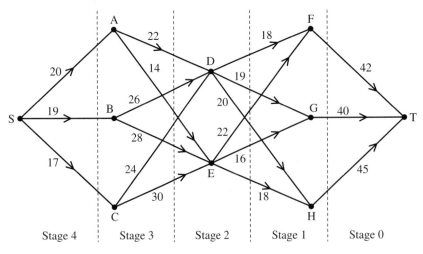

The network has been drawn so that the various stages are clear. The calculations are shown in the table below, where * indicates the minimum value when necessary.

Stage	Initial state	Action	Destination	Value
1	F	FT	T	42*
	G	GT	T	40*
	H	HT	T	45*
2	D	DF	F	$18 + 42 = 60$
		DG	G	$19 + 40 = 59*$
		DH	H	$20 + 45 = 65$
	E	EF	F	$22 + 42 = 64$
		EG	G	$16 + 40 = 56*$
		EH	H	$18 + 45 = 63$
3	A	AD	D	$22 + 59 = 81$
		AE	E	$14 + 56 = 70*$
	B	BD	D	$26 + 59 = 85$
		BE	E	$28 + 56 = 84*$
	C	CD	D	$24 + 59 = 83*$
		CE	E	$30 + 56 = 86$
4	S	SA	A	$20 + 70 = 90*$
		SB	B	$19 + 84 = 103$
		SC	C	$17 + 83 = 100$

Reading upwards through the table we see that the route that involves the minimum total cost is SAEGT.

The minimum total cost is 90.

Exercise 5C

1 An aircraft is required to fly from airport S to airport T. There are a number of routes possible. The airline wishes to choose the route for which the maximum amount of fuel used on any leg is as small as possible (the minimax route). The amounts of fuel used on the possible legs are shown in the table, in appropriate units.

Leg	Units
S to A	4
S to B	7
S to C	5
S to D	2
A to E	6
B to E	5
B to F	3
C to F	5
D to F	6
D to G	7
E to T	3
F to T	4
G to T	5

(a) Draw a network to show this information. Indicate the stages.

(b) Use dynamic programming to obtain the optimal route.

2

	Home (S)	A	B	C	D	E	F	G	H
Home (S)		30	40	45					
A					15	22			
B					18	24			
C					19	20			
D							25	20	21
E							27	19	22
F	36								
G	24								
H	20								

The table above shows the profits, in units of £10, a salesman estimates he will make on various journeys between some towns, one of which is S, where he lives (home). He wants to find the route that starts and finishes at S and is such that he makes the maximum total profit.

(a) Draw a network to model the above information with S as the start vertex and S as the destination vertex.

(b) On your network show the stages.

(c) Use dynamic programming to find the best route for the salesman.

3 A charity decides to support three groups, A, B and C, by giving all it raises each year to just one of them at a time. Over a 3-year period all the groups will receive aid. The directors of the charity decide that the policy to be adopted is to maximise the minimum return on its outlay. The benefits resulting from the various donations are estimated on a 10-point scale and shown in the table.

Already aided	Benefits		
	A	B	C
None	8	5	3
A	—	2	6
B	2	—	5
C	7	4	—
A and B	—	—	6
A and C	—	1	—
B and C	3	—	—

Find the policy that should be adopted.

SUMMARY OF KEY POINTS

1 **Bellman's principle for dynamic programming**: 'Any part of an optimal path is optimal'.

2 **A minimax** route is one on which the maximum length of edges used is as small as possible.

3 **A maximin** route is one for which the minimum length of the edges used is as large as possible.

Examination style paper 1

D2

1

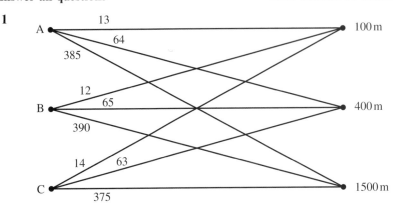

Downhill School has been invited to enter three pupils for an inter-school athletic competition. The competition consists of three track events: 100 m, 400 m and 1500 m. A pupil can only enter one event and the winning team is the one whose total time is a minimum. Downhill School has held trials to find the times, in seconds, taken by three pupils, A, B and C, in each of the events. These times are given in the diagram above. The teacher's problem is to decide who shall run which event to obtain the best result in the competition. Formulate the teacher's problem as a linear programming problem.   **(7 marks)**

2   An electrical shop has just received three new repair jobs: a television (1), a microwave (2) and a vacuum cleaner (3). Four men are available to do the repairs. The manager estimates what it will cost, in wages, to assign each of the workers to each of the jobs. These estimates, in £s, are given in the table below.

Jobs

Workers	1	2	3
Alf	14	16	11
Ben	13	15	12
Cyril	12	12	11
Dennis	16	18	16

Use the Hungarian algorithm to obtain the assignment of workers to jobs that results in the minimum overall cost. **(9 marks)**

3

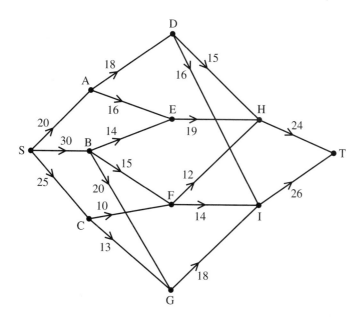

Members of a scout troup assemble at S and are told to make their way to T by any of the routes shown in the above figure. The weights on the edges give the time, in minutes, required to complete that leg. At least one person will take each of the possible routes from S to T.

Use dynamic programming to determine the maximum time that any scout could take to travel from S to T. State the route used. Show your calculations in tabular form. **(11 marks)**

4 Allan and Barbara play a zero-sum game and each has three possible strategies. The payoff matrix is shown below.

		B		
		I	**II**	**III**
	I	3	5	4
A	**II**	2	3	3
	III	6	3	8

(a) Explain why Allan will never choose strategy II. **(2 marks)**
(b) Explain why Barbara will never choose strategy III. **(2 marks)**
(c) Obtain the optimal strategies for both players and the value of the game. **(10 marks)**

5

Destination

		A	B	C	Available
	1	25	20	26	30
Source	**2**	15	20	20	100
	3	30	18	12	80
	Required	100	50	60	

The above table shows the unit transportation costs, the source availabilities and the destination requirements for a transportation problem. Use the north-west corner rule to obtain an initial basic feasible solution and then use the stepping-stone method to obtain the optimal solution. State the transportation pattern and give its cost. **(16 marks)**

6

	A	B	C	D	E	F
A	—	96	192	246	155	120
B	96	—	99	190	96	132
C	192	99	—	140	86	174
D	246	190	140	—	95	154
E	155	96	86	95	—	94
F	120	132	174	154	94	—

A sales representative, Bill, needs to visit stores in six cities, A, B, C, D, E and F. The table give the distances, in kilometres, between these six cities. Bill lives in city A and plans a route starting and finishing at A. He wishes to visit each city and drive the minimum distance.

(a) Starting from A, use Prim's algorithm to obtain a minimum spanning tree. State the order in which you selected the arcs and draw the tree. **(5 marks)**

(b) (i) Using your answer to (a) determine an initial upper bound for the length of the route planned by Bill.
(ii) Starting from your initial upper bound and using shortcuts obtain a route that is less than 725 km. **(6 marks)**

(c) By deleting city E determine a lower bound for Bill's route. **(4 marks)**

(d) From your calculations in part (c) deduce a route for Bill that starts at A, visits each city once, returns to A and involves a journey of less than 650 km. **(3 marks)**

Examination style paper 2

D2

1 The Coal Move Company specialises in coal handling. On Friday afternoon it has empty trucks at the following towns in the numbers indicated:

Town	Supply of Trucks
Ark Town (A)	35
Berryhill (B)	60
Creek Bottom (C)	25

On the following Monday morning the following towns will need the trucks shown:

Town	Demand for Trucks
River Halt (R)	30
Silver Sands (S)	45
Trout Pond (T)	45

Using a railway city-to-city distance chart, the dispatcher constructs a mileage table for the preceding towns. The result is:

	R	S	T
A	50	30	70
B	20	60	10
C	100	40	80

The company's aim is to move the trucks to new locations while minimising the total distance covered. Formulate this situation as a linear programming problem.   **(7 marks)**

2

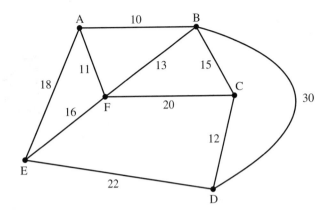

The figure above shows six villages and the roads connecting them. The lengths of the roads are given in kilometres.

(a) Complete the table below, in which the entries are the shortest distances between the villages. This may be done by inspection.

(2 marks)

	A	B	C	D	E	F
A	—	10	25		18	11
B	10	—	15			13
C	25	15	—	12		20
D			12	—	22	32
E	18			22	—	16
F	11	13	20	32	16	—

(b) Use the nearest neighbour algorithm, on the complete network modelled by using your completed table, to obtain an upper bound for the length of a tour in this network that starts and finishes at A and visits every village exactly once. **(3 marks)**

(c) Interpret your result in part (b) in terms of the original network. **(2 marks)**

(d) By choosing a different vertex as the starting point, use the nearest neighbour algorithm to obtain a shorter tour than that found in part (b). State the tour and the length of the tour.

(3 marks)

3

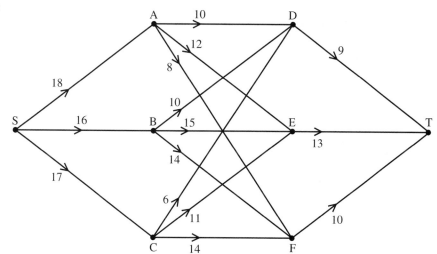

For the network shown above use dynamic programming to find a route from S to T such that the minimum length of arc used is as large as possible. That is find a maximin route. Show your calculations in tabular form.   **(11 marks)**

4 A detergent manufacturer has to decide how to schedule four one-minute commercials on four TV networks: channels A, B, C and D. One advert is to be in each of the time slots 1–2 p.m., 2–3 p.m., 3–4 p.m. and 4–5 p.m.

The exposure ratings for each hour are given in the table below.

Hours	**A**	**B**	**C**	**D**
1–2 p.m.	27	18	11	9
2–3 p.m.	19	15	17	10
3–4 p.m.	19	18	10	8
4–5 p.m.	11	21	17	13

Use the Hungarian algorithm to determine which channel should be scheduled each hour to provide **maximum** audience exposure.
    (13 marks)

5 Two players, A and B, play a zero-sum game. The payoff matrix for the game is:

B

A

−1	3	−9
2	−4	10

(*a*) Obtain, by graphical means, the optimal strategy of player A and the value of the game.   **(10 marks)**

(*b*) Hence obtain the optimal strategy of player B.   **(5 marks)**

6 A landscaping company has two new projects, A and B, each of which requires some infill material. Three sources, 1, 2 and 3, are able to supply infill material. The project requirements and the source availabilities, in truckloads, are given in the table below. The table also includes the shipping costs, in £10s, of the routes available.

Sources	A	B	Availability
1	8	10	45
2	10	12	18
3	11	10	17
Requirement	25	35	

The company wishes to keep its transportation costs to a minimum.

(*a*) Use the transportation algorithm to obtain a transportation pattern that has minimum cost. Find this cost. **(8 marks)**

(*b*) Comment on the implications of your solution. **(2 marks)**

(*c*) Is there another transportation pattern with this cost? If so find it. **(9 marks)**

Appendix 1: The dual of a linear programming problem

If you have studied the unit P6 you will have come across matrices and their properties. In section 4.6 the **transpose** of a matrix was referred to.

Definition:
If $\mathbf{A}$ is a matrix then the transpose of $\mathbf{A}$, written $\mathbf{A}^T$, is obtained from $\mathbf{A}$ by interchanging rows and columns.

Example 1

Given $\mathbf{A} = \begin{pmatrix} 8 & 7 & 1 \\ 6 & 6 & 9 \\ 5 & 7 & 7 \end{pmatrix}$

the transpose of $\mathbf{A}$ is

$\mathbf{A}^T = \begin{pmatrix} 8 & 6 & 5 \\ 7 & 6 & 7 \\ 1 & 9 & 7 \end{pmatrix}$

Example 2
If $\mathbf{x}$ is the column vector $\begin{pmatrix} x_1 \\ x_2 \\ x_3 \end{pmatrix}$ then the transpose of $\mathbf{x}$ is given

by $(x_1 \; x_2 \; x_3)$.

For example if $\mathbf{a} = \begin{pmatrix} 3 \\ 1 \\ 2 \end{pmatrix}$ then $\mathbf{a}^T = (3 \; 1 \; 2)$.

Consider the linear programming problem

Maximise $z = c_1 x_1 + c_2 x_2 + c_3 x_3$

subject to
$$a_{11}x_1 + a_{12}x_2 + a_{13}x_3 \leqslant b_1$$
$$a_{21}x_1 + a_{22}x_2 + a_{23}x_3 \leqslant b_2$$
$$a_{31}x_1 + a_{32}x_2 + a_{33}x_3 \leqslant b_3$$
$$x_1 \geqslant 0, \quad x_2 \geqslant 0, \quad x_3 \geqslant 0$$

Using matrix notation this may be written:

Maximise $\qquad z = \mathbf{c}^T \mathbf{x}$,

subject to $\qquad \mathbf{A}\mathbf{x} \leqslant \mathbf{b}$,

$\qquad\qquad\qquad \mathbf{x} \geqslant \mathbf{0}$,

where $\mathbf{x} = \begin{pmatrix} x_1 \\ x_2 \\ x_3 \end{pmatrix}$, $\mathbf{c} = \begin{pmatrix} c_1 \\ c_2 \\ c_3 \end{pmatrix}$,

$$\mathbf{A} = \begin{pmatrix} a_{11} & a_{12} & a_{13} \\ a_{21} & a_{22} & a_{23} \\ a_{31} & a_{32} & a_{33} \end{pmatrix} \text{ and } \mathbf{b} = \begin{pmatrix} b_1 \\ b_2 \\ b_3 \end{pmatrix}$$

This is usually called the **primal problem**.

The dual problem is:

Minimise $\qquad\qquad\qquad Z = \mathbf{b}^{\mathrm{T}}\mathbf{w}$

subject to $\qquad\qquad\qquad \mathbf{A}^{\mathrm{T}}\mathbf{w} \geqslant \mathbf{c}$

$$\mathbf{w} \geqslant \mathbf{0}$$

This is usually called the **dual problem**.

Example 3

Obtain the dual problem of the primal problem

maximise $\qquad\qquad\qquad z = 4x_1 + 10x_2 + 6x_3$

subject to
$$2x_1 + 4x_2 + x_3 \leqslant 12$$
$$6x_1 + 2x_2 + x_3 \leqslant 26$$
$$5x_1 + x_2 + 2x_3 \leqslant 80$$
$$x_1 \geqslant 0, \quad x_2 \geqslant 0, \quad x_3 \geqslant 0$$

Here $\mathbf{x} = \begin{pmatrix} x_1 \\ x_2 \\ x_3 \end{pmatrix}$ $\mathbf{c}^{\mathrm{T}} = (4\ 10\ 6)$

$$\mathbf{A} = \begin{pmatrix} 2 & 4 & 1 \\ 6 & 2 & 1 \\ 5 & 1 & 2 \end{pmatrix} \text{ and } \mathbf{b} = \begin{pmatrix} 12 \\ 26 \\ 80 \end{pmatrix}.$$

So $\mathbf{A}^{\mathrm{T}} = \begin{pmatrix} 2 & 6 & 5 \\ 4 & 2 & 1 \\ 1 & 1 & 2 \end{pmatrix}$

$$\mathbf{c} = \begin{pmatrix} 4 \\ 10 \\ 6 \end{pmatrix} \text{ and } \mathbf{b}^{\mathrm{T}} = (12\ 26\ 80)$$

Therefore the dual problem is

minimise $Z = (12\ 26\ 80) \begin{pmatrix} w_1 \\ w_2 \\ w_3 \end{pmatrix}$

subject to $\begin{pmatrix} 2 & 6 & 5 \\ 4 & 2 & 1 \\ 1 & 1 & 2 \end{pmatrix} \begin{pmatrix} w_1 \\ w_2 \\ w_3 \end{pmatrix} \geqslant \begin{pmatrix} 4 \\ 10 \\ 6 \end{pmatrix}$

that is

minimise $\qquad Z = 12w_1 + 26w_2 + 80w_3$

subject to $\qquad 2w_1 + 6w_2 + 5w_3 \geqslant 4$

$\qquad\qquad\qquad 4w_1 + 2w_2 + w_3 \geqslant 10$

$\qquad\qquad\qquad w_1 + w_2 + 2w_3 \geqslant 6$

$\qquad\quad w_1 \geqslant 0, \quad w_2 \geqslant 0, \quad w_3 \geqslant 0.$

Appendix 2: Other ways of formulating a game as a linear programming problem

In section 4.6 we showed how a game could be formulated as a linear programming problem, which could then be solved by the simplex method. In this appendix we discuss two other ways in which such a game can be formulated as a linear programming problem.

In both of these methods it is necessary for the value of the game to be non-negative because it appears as one of the decision variables. Recall that the simplex algorithm requires all decision variables to be non-negative. The first step is then, as before, to add a constant to each entry in the payoff matrix so that this condition is satisfied.

To make a direct comparison, consider the payoff matrix used in section 4.6, namely:

1	0	−6
−1	−1	2
−2	0	0

Again for direct comparison we add 7 to each entry and obtain:

	B_1	B_2	B_3
A_1	8	7	1
A_2	6	6	9
A_3	5	7	7

Let v be the value of the game as before. Let us consider the game from A's point of view. Suppose player A chooses strategies A_1, A_2 and A_3 in the proportions q_1, q_2 and q_3, then:

$$q_1 + q_2 + q_3 = 1$$

As before, we obtain:

$$8q_1 + 6q_2 + 5q_3 \geqslant v$$
$$7q_1 + 6q_2 + 7q_3 \geqslant v$$

$$q_1 + 9q_2 + 7q_3 \geqslant v$$
$$q_1 \geqslant 0, \quad q_2 \geqslant 0, \quad q_3 \geqslant 0$$

and v is to be **maximised**.

Method 1

In this method we replace the constraint:

$$q_1 + q_2 + q_3 = 1$$

by the weaker constraint:

$$q_1 + q_2 + q_3 \leqslant 1$$

This makes it easier to apply the simplex algorithm.

If we do this we obtain the linear programming problem:

Maximise v

subject to
$$v - 8q_1 - 6q_2 - 5q_3 \leqslant 0$$
$$v - 7q_1 - 6q_2 - 7q_3 \leqslant 0$$
$$v - q_1 - 9q_2 - 7q_3 \leqslant 0$$
$$q_1 + q_2 + q_3 \leqslant 1$$
$$v, q_1, q_2, q_3 \geqslant 0$$

Introducing slack variables r, s, t and u, the constraints can be written:

$$v - 8q_1 - 6q_2 - 5q_3 + r = 0$$
$$v - 7q_1 - 6q_2 - 7q_3 + s = 0$$
$$v - q_1 - 9q_2 - 7q_3 + t = 0$$
$$q_1 + q_2 + q_3 + u = 1$$

The simplex tableaux for solving this problem are:

Initial tableau

Basic Variable	v	q_1	q_2	q_3	r	s	t	u	Value
r	①	-8	-6	-5	1	0	0	0	0
s	1	-7	-6	-7	0	1	0	0	0
t	1	-1	-9	-7	0	0	1	0	0
u	0	1	1	1	0	0	0	1	1
P	-1	0	0	0	0	0	0	0	0

Taking $\textcircled{1}$ as the pivot:

Iteration 1

Basic Variable	v	q_1	q_2	q_3	r	s	t	u	Value
v	1	-8	-6	-5	1	0	0	0	0
s	0	$\textcircled{1}$	0	-2	-1	1	0	0	0
t	0	7	-3	-2	-1	0	1	0	0
u	0	1	1	1	0	0	0	1	1
P	0	-8	-6	-5	1	0	0	0	0

Taking $\textcircled{1}$ as the pivot:

Iteration 2

Basic Variable	v	q_1	q_2	q_3	r	s	t	u	Value
v	1	0	-6	-21	-7	8	0	0	0
q_1	0	1	0	-2	-1	1	0	0	0
t	0	0	-3	$\textcircled{12}$	6	-7	1	0	0
u	0	0	1	3	1	-1	0	1	1
P	0	0	-6	-21	-7	8	0	0	0

Taking $\textcircled{12}$ as the pivot:

Iteration 3

Basic Variable	v	q_1	q_2	q_3	r	s	t	u	Value
v	1	0	$-11\frac{1}{4}$	0	$3\frac{1}{2}$	$-4\frac{1}{4}$	$1\frac{3}{4}$	0	0
q_1	0	1	$-\frac{1}{2}$	0	0	$-\frac{1}{6}$	$\frac{1}{6}$	0	0
q_3	0	0	$-\frac{1}{4}$	1	$\frac{1}{2}$	$-\frac{7}{12}$	$\frac{1}{12}$	0	0
u	0	0	$\textcircled{$1\frac{3}{4}$}$	0	$-\frac{1}{2}$	$\frac{3}{4}$	$-\frac{1}{4}$	1	1
P	0	0	$-11\frac{1}{4}$	0	$3\frac{1}{2}$	$-4\frac{1}{4}$	$1\frac{3}{4}$	0	0

Taking $\textcircled{$1\frac{3}{4}$}$ as the pivot:

Iteration 4

Basic Variable	v	q_1	q_2	q_3	r	s	t	u	Value
v	1	0	0	0	$\frac{2}{7}$	$\frac{4}{7}$	$\frac{1}{7}$	$6\frac{3}{7}$	$6\frac{3}{7}$
q_1	0	1	0	0	$-\frac{1}{7}$	$\frac{1}{21}$	$\frac{2}{21}$	$\frac{2}{7}$	$\frac{2}{7}$
q_3	0	0	0	1	$\frac{3}{7}$	$-\frac{10}{21}$	$\frac{1}{21}$	$\frac{1}{7}$	$\frac{1}{7}$
q_2	1	0	1	0	$-\frac{2}{7}$	$\frac{3}{7}$	$-\frac{1}{7}$	$\frac{4}{7}$	$\frac{4}{7}$
P	0	0	0	0	$\frac{2}{7}$	$\frac{4}{7}$	$\frac{1}{7}$	$6\frac{3}{7}$	$6\frac{3}{7}$

This tableau is optional.

We obtain:

$$v = 6\tfrac{3}{7}$$
$$q_1 = \tfrac{2}{7}$$
$$q_2 = \tfrac{4}{7}$$
$$q_3 = \tfrac{1}{7}$$
$$P = 6\tfrac{3}{7}$$

These are, of course, the same as the results obtained in section 4.6.

As in section 4.6 we can also obtain from the final tableau the optimal strategy of the other player, in this case B. From the entries in the r, s and t columns and the final row we obtain B's optimal strategy:

play strategy B_1 for $\tfrac{2}{7}$ of the time
play strategy B_2 for $\tfrac{4}{7}$ of the time
play strategy B_3 for $\tfrac{1}{7}$ of the time

It should be noted that in this method we have four constraints, four decision variables (v, q_1, q_2 and q_3) and four slack variables (r, s, t and u). The number of individual calculations required is therefore greater than in the method given in section 4.6. However, the fractions involved in these calculations are often simpler.

Method 2

In this method we use the constraint: $q_1 + q_2 + q_3 = 1$ to eliminate one of the variables, for example q_3. We have $q_3 = 1 - q_1 - q_2$. Using this in the other constraints we have:

$$8q_1 + 6q_2 + 5q_3 = 8q_1 + 6q_2 + 5 - 5q_1 - 5q_2 = 3q_1 + q_2 + 5 \geqslant v$$

or $v - 3q_1 - q_2 \leqslant 5$

$$7q_1 + 6q_2 + 7q_3 = 7q_1 + 6q_2 + 7 - 7q_1 - 7q_2 = -q_2 + 7 \geqslant v$$

or $v + q_2 \leqslant 7$

$$q_1 + 9q_2 + 7q_3 = q_1 + 9q_2 + 7 - 7q_1 - 7q_2 = -6q_1 + 2q_2 \geqslant v$$

or $v + 6q_1 - 2q_2 \leqslant 7$

The linear programming problem is now:

Maximise v
subject to
$$v - 3q_1 - q_2 \leqslant 5$$
$$v + q_2 \leqslant 7$$
$$v + 6q_1 - 2q_2 \leqslant 7$$
$$v, q_1, q_2 \geqslant 0$$

Introducing slack variables R, S and T the constraints can be written:
$$v - 3q_1 - q_2 + R = 5$$
$$v + q_2 + S = 7$$
$$v + 6q_1 - 2q_2 + T = 7$$

The simplex tableaux for solving this problem are:

Initial tableau

Basic Variable	v	q_1	q_2	R	S	T	Value
R	①	-3	-1	1	0	0	5
S	1	0	1	0	1	0	7
T	1	6	-2	0	0	1	7
P	-1	0	0	0	0	0	0

Taking ① as the pivot:

Iteration 1

Basic Variable	v	q_1	q_2	R	S	T	Value
v	1	-3	-1	1	0	0	5
S	0	3	2	-1	1	0	2
T	0	⑨	-1	-1	0	1	2
P	0	-3	-1	1	0	0	5

Taking ⑨ as the pivot:

Iteration 2

Basic Variable	v	q_1	q_2	R	S	T	Value
v	1	0	$-1\frac{1}{3}$	$\frac{2}{3}$	0	$\frac{1}{3}$	$5\frac{2}{3}$
S	0	0	②$\frac{1}{3}$	$-\frac{2}{3}$	1	$-\frac{1}{3}$	$1\frac{1}{3}$
q_1	0	1	$-\frac{1}{9}$	$-\frac{1}{9}$	0	$\frac{1}{9}$	$\frac{2}{9}$
P	0	0	$-1\frac{1}{3}$	$\frac{2}{3}$	0	$\frac{1}{3}$	$5\frac{2}{3}$

Taking ②$\frac{1}{3}$ as the pivot:

Iteration 3

Basic Variable	v	q_1	q_2	R	S	T	Value
v	1	0	0	$\frac{2}{7}$	$\frac{4}{7}$	$\frac{1}{7}$	$6\frac{3}{7}$
q_2	0	0	1	$-\frac{2}{7}$	$\frac{3}{7}$	$-\frac{1}{7}$	$\frac{4}{7}$
q_1	0	1	0	$-\frac{1}{7}$	$\frac{1}{21}$	$\frac{2}{21}$	$\frac{2}{7}$
P	0	0	0	$\frac{2}{7}$	$\frac{4}{7}$	$\frac{1}{7}$	$6\frac{3}{7}$

This tableau is optimal.

We obtain
$$q_1 = \tfrac{2}{7}$$
$$q_2 = \tfrac{4}{7}$$

So
$$q_3 = 1 - \tfrac{2}{7} - \tfrac{4}{7} = \tfrac{1}{7}$$
$$P = v = 6\tfrac{3}{7}$$

The results obtained are again the same as those obtained in section 4.6. The optimal strategy for player B is obtained from the entries in the R, S and T columns and the bottom row of the optimal tableau.

The number of decision variables involved in this method is three and the number of slack variables is three as there are now only three constraints. In addition, one of the original decision variables has to be eliminated. Depending on the payoff matrix, the fractions involved may or may not be simpler.

Answers

Exercise 1A

1 (a)

To From	C_1	C_2	C_3	Supply
M_1	11	9	3	40
M_2	7	3	5	10
M_3	9	3	3	20
Demand	12	18	40	

(b) This is a balanced problem.

Let x_{ij} be the number of kilotonnes shipped from M_i to C_j ($i = 1, 2, 3$ and $j = 1, 2, 3$). Total cost is Z thousand pounds.

Minimise $Z = 11x_{11} + 9x_{12} + 3x_{13}$
$$+7x_{21} + 3x_{22} + 5x_{23}$$
$$+9x_{31} + 3x_{32} + 3x_{33}$$

subject to $\left.\begin{array}{l} x_{11} + x_{12} + x_{13} = 40 \\ x_{21} + x_{22} + x_{23} = 10 \\ x_{31} + x_{32} + x_{33} = 20 \end{array}\right\}$ rows

$\left.\begin{array}{l} x_{11} + x_{21} + x_{31} = 12 \\ x_{12} + x_{22} + x_{32} = 18 \\ x_{13} + x_{23} + x_{33} = 40 \end{array}\right\}$ columns

Non-negativity condition $x_{ij} \geqslant 0$ for $i = 1, 2, 3$ and $j = 1, 2, 3$

2 This is a balanced problem. Let x_{ij} be the number transported from F_i to W_j. Z is the total cost of transportation.

Minimise $Z = 7x_{11} + 8x_{12} + 6x_{13}$
$$+9x_{21} + 2x_{22} + 4x_{23}$$
$$+5x_{31} + 6x_{32} + 3x_{33}$$

subject to $x_{11} + x_{12} + x_{13} = 4$
$$x_{21} + x_{22} + x_{23} = 3$$
$$x_{31} + x_{32} + x_{33} = 8$$
$$x_{11} + x_{21} + x_{31} = 2$$

$$x_{12} + x_{22} + x_{32} = 9$$
$$x_{13} + x_{23} + x_{33} = 4$$

$x_{ij} \geqslant 0$ for $i = 1, 2, 3$ and $j = 1, 2, 3$

3 This is a balanced problem.

Let x_{ij} be the number of tons of carpet shipped from plant i to outlet j. Let £Z be the cost of transportation.

Minimise $Z = 20x_{AC} + 32x_{AD}$
$$+35x_{BC} + 15x_{BD}$$

subject to $x_{AC} + x_{AD} = 250$
$$x_{BC} + x_{BD} = 300$$
$$x_{AC} + x_{BC} = 320$$
$$x_{AD} + x_{BD} = 230$$

$x_{ij} \geqslant 0$ for i $=$A, B, j $=$ C, D

Exercise 1B

1 (a)

60	40	
	20	20
		50

(b)

60		
40		
20	80	40

(c)

60		
	70	30
		50

(d)

60	30	
	40	
		50

2 (a)

90	10
	150

(b)

	100
90	60

(c)

	100
90	60

This is the same as (b)

(d)

90	10
	150

This is the same as (a)

Cost of (a): $(90 \times 8) + (10 \times 6) + (150 \times 12)$
$= 2580$

Cost of (b): $(100 \times 6) + (90 \times 10) + (60 \times 12)$
$= 2220$

Hence (b) is the optimal solution.

$1 \to$ B send 100

$2 \to$ A send 90

$2 \to$ B send 60

Cost 2220

Exercise 1C

1 $I_{12} = 1$, $I_{13} = 0$, $I_{23} = 4$, $I_{31} = 4$

All these values are non-negative so the solution is optimal.

$1 \to$ A send 40, $2 \to$ A send 30, $2 \to$ B send 20,

$3 \to$ B send 30, $3 \to$ C send 30

Cost $= 1040$

2 (a)

4		
3		
2	4	2

cost $= 72$

(b) $I_{12} = 1$, $I_{13} = 0$, $I_{22} = 5$, $I_{23} = 8$

All values are non-negative and so the solution is optimal.

3 (a)

8	6	
	4	6
		6

(b) $I_{13} = 4$, $I_{21} = 1$, $I_{31} = -1$, $I_{32} = 0$

Since I_{31} is negative the solution is not optimal and so can be improved.

Exercise 1D

1 The improvement indices obtained were

$I_{13} = 4$, $I_{21} = 1$, $I_{31} = -1$ and $I_{32} = 0$.

The loop required to use (3, 1) is

X	X	
	X	X
		X

Adding and subtracting θ as appropriate gives:

$8 - \theta$	$6 + \theta$	
	$4 - \theta$	$6 + \theta$
$+\theta$		$6 - \theta$

Take $\theta = 4$ and obtain the improved solution:

4	10	
		10
4		2

The improvement indices are $I_{13} = 3$, $I_{21} = 2$, $I_{22} = 1$ and $I_{32} = 3$.

Since all are positive this solution is optimal:

send 4 units from $S_1 \to W_1$

10 units from $S_1 \to W_2$

10 units from $S_2 \to W_3$

4 units from $S_3 \to W_1$

2 units from $S_3 \to W_3$

Total minimum cost $= (4 \times 10) + (10 \times 4)$
$+ (10 \times 8) + (4 \times 9) + (2 \times 7) = 210$

2 North-west corner solution

25		
5	30	5
		31

Improved solution

25		
	30	10
5		26

Optimal solution

		25
	30	10
30		1

Total cost = 233
$A_1 \rightarrow B_3$ send 25
$A_2 \rightarrow B_2$ send 30
$A_2 \rightarrow B_3$ send 10
$A_3 \rightarrow B_1$ send 30
$A_3 \rightarrow B_3$ send 1

Exercise 1E

1 (a) The north-west corner solution of the corresponding balanced problem is:

From \ To	R_1	R_2	Dummy	Supply
F_1	30			30
F_2	30	20	10	60
F_3			20	20
Demand	60	20	30	110

(b) The improvement indices are $I_{12} = 2$, $I_{13} = 0$, $I_{31} = 4$ and $I_{32} = 5$, so the solution is optimal.

(c) Send 30 units from $F_1 \rightarrow R_1$
  30 units from $F_2 \rightarrow R_1$
  20 units from $F_2 \rightarrow R_2$
10 units remain at F_2 and 20 units remain at F_3
The total cost of this pattern is 200

2 (a) Total demand = 70 + 30 + 40 = 140
  Total supply = 50 + 60 = 110
Since total demand $\neq$ total supply, problem is unbalanced.

From \ To	D_1	D_2	D_3	Supply
S_1	7	8	10	50
S_2	9	7	8	60
Dummy	0	0	0	30
Demand	70	30	40	140

(b) The north-west corner solution is:

50		
20	30	10
		30

Use of the stepping-stone method once produces the optimal solution:

	D_1	D_2	D_3
S_1	50		
S_2		30	30
Dummy	20		10

(c) Send 50 units from S_1 to D_1
  30 units from S_2 to D_2
  30 units from S_2 to D_3
D_1 does not receive 20 units it wants
D_3 does not receive 10 units it wants
Total minimum cost is 800

Exercise 1F

1 (a)

25		
5	30	
		20

There are only four occupied cells, which is less than 3 + 3 − 1 = 5

(b) Optimal solution:

	3	2	3
0	25	0	0
0	1	15	20
0	5	15	3

(c) $C_1 \rightarrow T_1$ (25 engines)
  $C_2 \rightarrow T_2$ (15 engines)
  $C_2 \rightarrow T_3$ (20 engines)
  $C_3 \rightarrow T_1$ (5 engines)
  $C_3 \rightarrow T_2$ (15 engines)
Cost = 210 × £100s

2 (a)

	A	B	C	
X	9	4		13
Y		9	2	11
Z			10	10
	9	13	12	

(b) Optimal solution:

	13	
9		2
		10

Minimum total cost = 93 (£10s)

3 (a)

10	20−θ	θ
	θ	20−θ
		10

(b) The next tableau is (e.g.):

10		20
	20	
		10

The optimal tableau is:

	10	20
10	10	
		10

Minimum cost = 710

Exercise 2A

1 Let $x_{ij} = \begin{cases} 1 & \text{if machine } i \text{ is assigned to} \\ & \text{component } j \\ 0 & \text{otherwise} \end{cases}$

Min $Z = 13x_{A1} + 14x_{A2} + 10x_{A3} + 11x_{B1}$
$\qquad + 16x_{B2} + 15x_{B3} + 17x_{C1}$
$\qquad + 12x_{C2} + 9x_{C3}$

subject to $x_{A1} + x_{A2} + x_{A3} = 1$
$\qquad x_{B1} + x_{B2} + x_{B3} = 1$
$\qquad x_{C1} + x_{C2} + x_{C3} = 1$
$\qquad x_{A1} + x_{B1} + x_{C1} = 1$
$\qquad x_{A2} + x_{B2} + x_{C2} = 1$
$\qquad x_{A3} + x_{B3} + x_{C3} = 1$

Exercise 2B

1 (a) The opportunity cost matrix is:

3	1	0*
0*	2	4
8	0*	0

(b) The minimum number of lines needed to cover the zeros is three, as shown. Hence an optimal assignment can be made.

(c) $\quad$ A → 3 cost 10
$\qquad$ B → 1 cost 11
$\qquad$ C → 2 cost 12

Total cost = 33

2 (a) The opportunity cost matrix is:

16	8	0
0	0	4
7	2	0

(b) The zeros can be covered by two lines and so no optimal assignment is possible.

3 (a) The opportunity cost matrix is:

4	0	1	2
0	0	13	0
18	0	1	17
14	0	0	16

(b) The zeros can be covered by three lines and so no optimal assignment is possible.

Exercise 2C

1 Revising the opportunity cost matrix gives

14	6	0
0	0	6
5	0	0

Three lines are required to cover the zeros. The optimal assignment is:

1 → C cost 10

2 → A cost 16

3 → B cost 14

Total minimum cost = 40

2 Revised opportunity matrix:

3	0	0	1
0	1	13	0
17	0	0	16
14	1	0	16

The zeros may be covered by three lines and so an optimal assignment is not possible.

Revising the opportunity matrix again gives:

2	0	0	0
0	2	14	0
16	0	0	15
13	1	0	15

A minimum number of four lines is required to cover the zeros, as shown. An optimal assignment is:

$A \rightarrow W$ cost 25

$B \rightarrow N$ cost 26

$C \rightarrow S$ cost 21

$D \rightarrow E$ cost 23

Total cost = 95

3 An optimal assignment is:

Bob $\rightarrow$ 1 cost 10

Sue $\rightarrow$ 3 cost 4

Jim $\rightarrow$ 2 cost 9

Amy $\rightarrow$ 4 cost 10

Total minimum cost = 33

Exercise 2D

1 $1 \rightarrow B, 2 \rightarrow D, 3 \rightarrow A$

Total time = 8

2 (i)

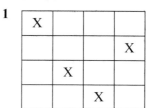

(ii)

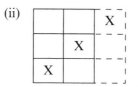

Minimum total cost = 11

Exercise 2E

1

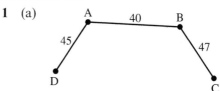

Total evaluation score = 330

Exercise 3A

1 (a)

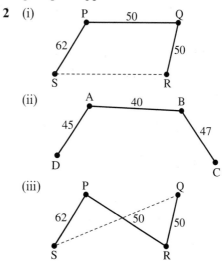

length 132

Hence an upper bound is

$2 \times 132 = 264$ for DABCBAD.

(b) If on reaching C we go directly to D we have a reduction of

$$(45 + 40 + 47) - 90(CD) = 42$$

giving an upper bound of 222.

2 (i)

(ii)

(iii)

Upper bound is $2 \times 162 = 324$

(i) Using RS gives

$$62 + 50 + 50 + RS(70) = 232$$

(ii) Using QR and SR also gives 232

(iii) Using only QS gives

$$62 + 50 + 50 + 104 = 266$$

3 (a) (i)

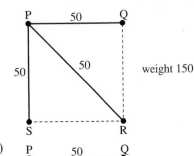

weight 150

(ii)

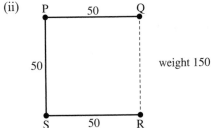

weight 150

(iii)

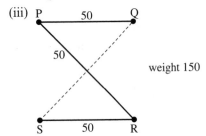

weight 150

Upper bound $2 \times 150 = 300$

(b) (i) Using shortcut shown 217

(ii) Using shortcut shown 217

(iii) Using shortcut shown 235

4 (a) Two possible minimum spanning trees:

(i)

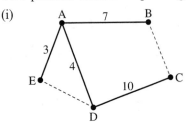

(ii)

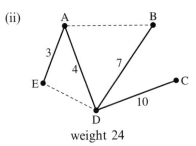

weight 24

Hence an upper bound is $2 \times 24 = 48$

(b) (i) Using shortcuts BC and ED in (i) reduces this by $(21 - 17) = 4$ and $(7 - 5) = 2$ and so gives 42.

(ii) Using BC does not improve upper bound but using AB does by $(11 - 7) = 4$ and ED as before reduces by 2, again giving 42.

5 (a)

	1	**4**	**3**	**2**	**5**
	A	**B**	**C**	**D**	**E**
A	—	17	10	9	12
B	17	—	⑧	14	5
C	10	8	—	⑦	11
D	⑨	14	7	—	11
E	12	⑤	11	11	—

AD (9), DC (7), CB (8), BE (5), length 29

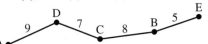

(b) Hence an upper bound is $2 \times 29 = 58$ (ADCBEBCDA).

(c) If when we reach E we shortcut back to A, the length is

$$9 + 7 + 8 + 5 + AE(12) = 41$$

which is a better upper bound.

6 (a) A minimum spanning tree is

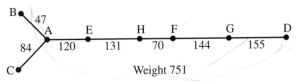

Weight 751

(b) An upper bound is
$2 \times$ weight of minimum spanning tree $= 1502$
ABACAEHFGDGFHEAA

(c)   E.g. if when we reach D we go directly to B we save

$$(47 + 120 + 131 + 70 + 144 + 155)$$
$$-BD\,(402) = 265$$

This reduces the upper bound to 1237 miles.

Exercise 3B

1   (a)   Weight of MST$(124) + 40 + 45 = 209$
(b)   Weight of MST$(120) + 40 + 47 = 207$
(c)   Weight of MST$(85) + 47 + 75 = 207$
(d)   Weight of MST$(87) + 45 + 77 = 209$

2   (a)   Weight of MST$(112) + 50 + 50 = 212$
(b)   Weight of MST$(100) + 62 + 70 = 232$
(c)   From Exercise 3A, question 2: length of tour $\leqslant 232$

From (a) and (b): length of tour $\geqslant 232$

$\therefore$ length of minimum tour $= 232$

From (b) this is PQRSP.

3   (a)   Weight of MST$(22) + 3 + 4 = 29$
(b)   Weight of MST$(17) + 7 + 7 = 31$
(c)   Weight of MST$(14) + 10 + 13 = 37$
(d)   Weight of MST$(23) + 5 + 4 = 32$
(e)   Weight of MST$(21) + 5 + 3 = 29$
(f) $37 \leqslant L \leqslant 42$
(g)   (e.g.) ABCDEA; $7 + 17 + 10 + 5 + 3 = 42$

4   (a)   Weight of MST$(20) + 9 + 10 = 39$
(b)   Weight of MST$(27) + 8 + 5 = 40$
(c)   Weight of MST$(25) + 8 + 7 = 40$
(d)   Weight of MST$(23) + 9 + 7 = 39$
(e)   Weight of MST$(24) + 5 + 11 = 40$
(f)   From (e) and answer to question 5 of Exercise 3A: $40 \leqslant L \leqslant 41$

In (e) we have

+EB(5) and either EC or ED(11).
EB joins vertex E to one end of the chain.
Neither EC nor ED does, but EA does. We then have EADCBE of length
$12 + 9 + 7 + 8 + 5 = 41.$

Hence a minimum tour is ADCBEA.

5   The MST found previously was:

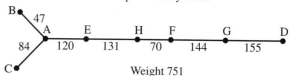

Weight 751

of weight 751.

(a)   If we delete B then the MST is obtained by removing edge AB from above, giving an MST of weight 704.
The two edges of smallest weight at B are BA(47) and BC(121).
Lower bound is $704 + 47 + 121 = 872$
(b)   As above,
(weight of MST) $= 751 - 84 = 667$
Weight of edges to be added $=$ CA(84) $+$ CB(121)
Lower bound is $667 + 84 + 121 = 872$
(c)   As above,
(weight of MST) $= 751 - 155 = 596$
Weight of edges to be added $=$ DF(220) $+$ DG(155)
Lower bound is $596 + 220 + 155 = 971$
(d) L satisfies $971 \leqslant L \leqslant 1237$

6   (a)   Best upper bound 53
(b)   Best lower bound 52 (removing A or D)
(c) $52 \leqslant L \leqslant 53$
(d)   Shortest possible route: ABEFCDA, length 53

Exercise 3C

1   (a)   A $\xrightarrow{3}$ E $\xrightarrow{5}$ D $\xrightarrow{7}$ B $\xrightarrow{17}$ C
13

AEDBCA length 45 units

(b)   B $\xrightarrow{7}$ A $\xrightarrow{3}$ E $\xrightarrow{5}$ D $\xrightarrow{10}$ C
17

BAEDCB length 42 units
or BDAECB length 44 units

2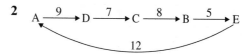

length 41

Exercise 3D

1 (a)

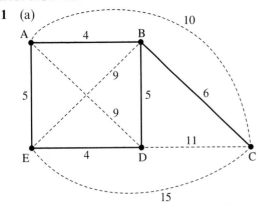

(b)

$$A \xrightarrow{4} B \xrightarrow{5} D \xrightarrow{4} E \xrightarrow{15} C$$

with return arc 10 from C to A

Tour ABDECA

Length 38

(c) Replace EC by ED, DB and BC

Replace CA by CB and BA

This gives ABDEDBCBA

2 (a)

	A	B	C	D	E
A	—	10	9	20	8
B	10	—	12	18	6
C	9	12	—	12	11
D	20	18	12	—	12
E	8	6	11	12	—

(b)

$$A \xrightarrow{8} E \xrightarrow{6} B \xrightarrow{12} C \xrightarrow{12} D$$

with return arc 20

Tour AEBCD length 58

In the original network replace DA by DE and EA, giving AEBCDEA.

(c) (i) Starting at D: DEBACD (49)

(ii) Starting at B: BEACDEB (53)

(iii) Starting at C: CAEBEDC (53)

(iv) Starting at E: EBACDE (49)

3

	A	B	C	D
A	—	4	7	6
B	4	—	3	8
C	7	3	—	5
D	6	8	5	—

Exercise 4A

1 (a) Max(row min) = 2, min(col max) = 4

(b) Max(row min) = 4, min(col max) = 5

(c) Max(row min) = 1, min(col max) = 2

(d) Max(row min) = 5, min(col max) = 6

2 (i) (a) Max(row min) = 4, min(col max) = 4

(b) row 1, col 2

(c) value of game = 4

(ii) (a) Max(row min) = −1, min(col max) = −1

(b) row 2, col 1

(c) value of game = −1

(iii) (a) Max(row min) = 0, min(col max) = 0

(b) row 3, col 2

(c) value of game = 0

(iv) (a) Max(row min) = 2, min(col max) = 2

(b) row 1, col 2

(c) value of game = 2

(v) (a) Max(row min) = 4, min(col max) = 4

(b) row 2, col 1

(c) value of game = 4

Exercise 4B

1 (a) (i) row 1 for $\frac{2}{5}$ of time

row 2 for $\frac{3}{5}$ of time

column 1 for $\frac{7}{10}$ of time

column 2 for $\frac{3}{10}$ of time

(ii) $5\frac{1}{5}$

(b) (i) row 1 for $\frac{1}{3}$ of time

row 2 for $\frac{2}{3}$ of time

column 1 for $\frac{3}{5}$ of time

column 2 for $\frac{2}{5}$ of time

(ii) 9

(c) (i) row 1 for $\frac{2}{3}$ of time

row 2 for $\frac{1}{3}$ of time

column 1 for $\frac{8}{21}$ of time

column 2 for $\frac{13}{21}$ of time

(ii) $\frac{2}{3}$

Exercise 4C

1 (a) row 1 for $\frac{3}{7}$ of time

row 2 for $\frac{4}{7}$ of time

column 1 for $\frac{3}{7}$ of time

column 2 for $\frac{4}{7}$ of time

do not use column 3 at all

Value $= 2\frac{5}{7}$

(b) row 1 for $\frac{3}{5}$ of time

row 2 for $\frac{2}{5}$ of time

do not use row 3 at all

column 1 for $\frac{3}{5}$ of time

column 2 for $\frac{2}{5}$ of time

Value $= -0.2$

Exercise 4D

1 (a) Add 4 to each entry to make all entries positive to obtain:

8	3	1
2	4	7

Linear programming problem to be solved:

Maximise $P = x_1 + x_2 + x_3$

subject to $8x_1 + 3x_2 + x_3 \leqslant 1$

$2x_1 + 4x_2 + 7x_3 \leqslant 1$

$x_1 \geqslant 0, x_2 \geqslant 0, x_3 \geqslant 0$

(b) $x_1 = \frac{1}{26}, x_2 = \frac{3}{13}, x_3 = 0$

$P = \frac{7}{26}$

So $v = \frac{26}{7} = 3\frac{5}{7}$

$p_1 = \frac{1}{7}, p_2 = \frac{6}{7}, p_3 = 0$

The value of the original game $= 3\frac{5}{7} - 4 = -\frac{2}{7}$

B's optimal strategy is:

use I for $\frac{1}{7}$ of the time

use II for $\frac{6}{7}$ of the time

do not use III at all

(c) From the final tableau:

$y_1 = \frac{1}{13}, y_2 = \frac{5}{26}, v = \frac{26}{7}$

So $q_1 = \frac{2}{7}, q_2 = \frac{5}{7}$

A's optimal strategy is:

use I for $\frac{2}{7}$ of the time

use II for $\frac{5}{7}$ of the time

2 (a) Maximise $P = x_1 + x_2 + x_3$

subject to $8x_1 + 4x_2 + 2x_3 \leqslant 1$

$2x_1 + 8x_2 + 4x_3 \leqslant 1$

$2x_1 + x_2 + 8x_3 \leqslant 1$

$x_1 \geqslant 0, x_2 \geqslant 0, x_3 \geqslant 0$

(b) $x_1 = \frac{1}{14}, x_2 = \frac{2}{35}, x_3 = \frac{1}{10}, P = \frac{8}{35}$

So $v = 4\frac{3}{8}$

$p_1 = \frac{5}{16}, p_2 = \frac{1}{4}, p_3 = \frac{7}{16}$

B's optimal strategy is:

use I for $\frac{5}{16}$ of the time

use II for $\frac{1}{4}$ of the time

use III for $\frac{7}{16}$ of the time

(c) From the final tableau:

$y_1 = \frac{19}{210}, y_2 = \frac{1}{14}, y_3 = \frac{1}{15}, Q = \frac{8}{35}$

So $v = 4\frac{3}{8}$

$q_1 = \frac{19}{48}, q_2 = \frac{5}{16}, q_3 = \frac{7}{24}$

A's optimal strategy is:

use I for $\frac{19}{48}$ of the time

use II for $\frac{5}{16}$ of the time

use III for $\frac{7}{24}$ of the time

Exercise 4E

1 (a) The payoff table can be reduced to

1.8	1.1
1.3	1.5

(b) A's optimal strategy is:

use I for $\frac{2}{9}$ of the time

use II for $\frac{7}{9}$ of the time

never use III

(c) B's optimal strategy is:

use I for $\frac{4}{9}$ of the time

use II for $\frac{5}{9}$ of the time

never use III

(d) value $= 1.411$

2 (a) The payoff table can be reduced to:

-2	4
0	-3

(b) X's optimal strategy is:

use I for $\frac{1}{3}$ of the time

use II for $\frac{2}{3}$ of the time

never use III

(c) Y's optimal strategy is:

use I for $\frac{7}{9}$ of the time

use II for $\frac{2}{9}$ of the time

never use III

(d) The value of game $= -\frac{2}{3}$

3 (a) Payoff table can be reduced to:

2	4
3	0
−1	6

(b) Using graphical methods, Q's optimal strategy is:

use I for $\frac{4}{5}$ of the time

use II for $\frac{1}{5}$ of the time

(c) P's optimal strategy:

use I for $\frac{3}{5}$ of the time

never use II

use III for $\frac{2}{5}$ of the time

never use IV

(d) The value of the game $= 2\frac{2}{5}$

Exercise 5A

1 (a) (i)

Stage	Initial state	Action	Destination	Value
1	C D E	CT DT ET	T T T	5* 4* 7*
2	A	AC AE	C E	11 + 5 = 16 6 + 7 = 13*
	B	BC BD BE	C D E	8 + 5 = 13 7 + 4 = 11* 5 + 7 = 12
3	S	SA SB	A B	8 + 13 = 21* 12 + 11 = 23

Shortest route is SAET, length 21

(ii)

Stage	Initial state	Action	Destination	Value
1	C D E	CT DT ET	T T T	5* 4* 7*
2	A	AC AE	C E	11 + 5 = 16* 6 + 7 = 13
	B	BC BD BE	C D E	8 + 5 = 13* 7 + 4 = 11 5 + 7 = 12
3	S	SA SB	A B	8 + 16 = 24 12 + 13 = 25*

Longest route is SBCT, length 25

(b) (i)

Stage	Initial state	Action	Destination	Value
1	F G H I	FT GT HT IT	T T T T	21* 22* 19* 20*
2	D	DF DH	F H	14 + 21 = 35 13 + 19 = 32*
	E	EF EG EH EI	F G H I	8 + 21 = 29 10 + 22 = 32 12 + 19 = 31 8 + 20 = 28*
3	A	AD AE	D E	8 + 32 = 40 9 + 28 = 37*
	B	BD BE	D E	6 + 32 = 38 8 + 28 = 36*
	C	CD CE	D E	10 + 32 = 42 7 + 28 = 35*
4	S	SA SB SC	A B C	20 + 37 = 57 15 + 36 = 51* 17 + 35 = 52

Shortest route is SBEIT, length 51

(ii)

Stage	Initial state	Action	Destination	Value
1	F G H I	FT GT HT IT	T T T T	21* 22* 19* 20*
2	D	DF DH	F H	$14 + 21 = 35$* $13 + 19 = 32$
	E	EF EG EH EI	F G H I	$8 + 21 = 29$ $10 + 22 = 32$* $12 + 19 = 31$ $8 + 20 = 28$
3	A	AD AE	D E	$8 + 35 = 43$* $9 + 32 = 41$
	B	BD BE	D E	$6 + 35 = 41$* $8 + 32 = 40$
	C	CD CE	D E	$10 + 35 = 45$* $7 + 32 = 39$
4	S	SA SB SC	A B C	$20 + 43 = 63$* $15 + 41 = 56$ $17 + 45 = 62$

Longest route is SADFT, length 63.

Exercise 5B

1 (a)

Stage	Initial state	Action	Destination	Value
1	C D E	CT DT ET	T T T	5* 4* 7*
2	A	AC AE	C E	$\max(11, 5) = 11$ $\max(6, 7) = 7$*
	B	BC BD BE	C D E	$\max(8, 5) = 8$ $\max(7, 4) = 7$* $\max(5, 7) = 7$*
3	S	SA SB	A B	$\max(8, 7) = 8$* $\max(12, 7) = 12$

* indicates minimum

Minimax route is SAET. The maximum on this route is 8.

(b)

Stage	Initial state	Action	Destination	Value
1	C D E	CT DT ET	T T T	5* 4* 7*
2	A	AC AE	C E	$\min(11, 5) = 5$ $\min(6, 7) = 6$*
	B	BC BD BE	C D E	$\min(8, 5) = 5$* $\min(7, 4) = 4$ $\min(5, 7) = 5$*
3	S	SA SB	A B	$\min(8, 6) = 6$* $\min(12, 5) = 5$

*indicates maximum

Maximin route is SAET. The minimum on this route is 6.

Exercise 5C

1

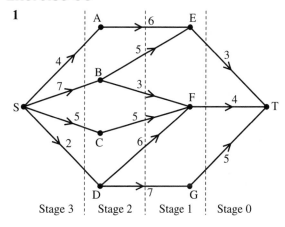

Stage 3 ┊ Stage 2 ┊ Stage 1 ┊ Stage 0

Stage	Initial state	Action	Destination	Value
1	E	ET	T	3*
	F	FT	T	4*
	G	GT	T	5*
2	A	AE	E	max(6, 3) = 6*
	B	BE	E	max(5, 3) = 5
		BF	F	max(3, 4) = 4*
	C	CF	F	max(5, 4) = 5*
	D	DF	F	max(6, 4) = 6*
		DG	G	max(7, 5) = 7
3	S	SA	A	max(4, 6) = 6
		SB	B	max(7, 4) = 7
		SC	C	max(5, 5) = 5*
		SD	D	max(2, 6) = 6

*indicates minimum

The required route is SCFT. Maximum amount used on any leg is 5.

2 (a), (b)

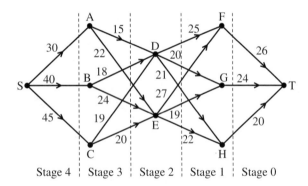

Stage 4 ┊ Stage 3 ┊ Stage 2 ┊ Stage 1 ┊ Stage 0

(c)

Stage	Initial state	Action	Destination	Value
1	F	FS	S	26*
	G	GS	S	24*
	H	HS	S	20*
2	D	DF	F	25 + 26 = 51*
		DG	G	20 + 24 = 44
		DH	H	21 + 20 = 41
	E	EF	F	27 + 26 = 53*
		EG	G	19 + 24 = 43
		EH	H	22 + 20 = 42
3	A	AD	D	15 + 51 = 66
		AE	E	22 + 53 = 75*
	B	BD	D	18 + 51 = 69
		BE	E	24 + 53 = 77*
	C	CD	D	19 + 51 = 70
		CE	E	20 + 53 = 73*
4	S	SA	A	30 + 75 = 105
		SB	B	40 + 77 = 117
		SC	C	45 + 73 = 118*

The best route is SCEFS.

The maximum profit is 118 × £10.

3

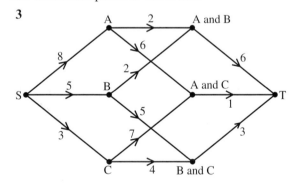

Stage	State	Destination	Value
1	A + B A + C B + C	T T T	6* 1* 3*
2	A	A + B A + C	min(2, 6) = 2* min(6, 1) = 1
	B	B + A B + C	min(2, 6) = 2 min(5, 3) = 3*
	C	C + A C + B	min(7, 1) = 1 min(4, 3) = 3*
3	S	A B C	min(8, 2) = 2 min(5, 3) = 3*(i) min(3, 3) = 3*(ii)

(i)   gives BCA

(ii) gives CBA

The minimum return is 3

Examination style paper 1

1

	100 m (1)	400 m (2)	1500 m (3)
A(1)	13	64	385
B(2)	12	65	390
C(3)	14	63	375

$x_{ij} = 1$ if pupil i is given event j

$x_{ij} = 0$ otherwise.

Total time $= Z$ seconds.

Minimise $Z = 13x_{11} + 64x_{12} + 385x_{13}$
$$+ 12x_{21} + 65x_{22} + 390x_{23}$$
$$+ 14x_{31} + 63x_{32} + 375x_{33}$$

Subject to $x_{11} + x_{12} + x_{13} = 1$ (A)

$x_{21} + x_{22} + x_{23} = 1$ (B)

$x_{31} + x_{32} + x_{33} = 1$ (C)

$x_{11} + x_{21} + x_{31} = 1$ (100 m)

$x_{12} + x_{22} + x_{32} = 1$ (400 m)

$x_{13} + x_{23} + x_{33} = 1$ (1500 m)

2 $A \to 3$

$B \to 1$

$C \to 2$

$D \to 4$ (i.e. unused)

Minimum overall cost = £36

3   Maximum time = 94 minutes

Route used = SBGIT

4   (a)   Each entry in row II is less than the corresponding entry in row I. Allan can therefore always do better by playing strategy I.

(b)   Each entry in column III is greater than the corresponding entry in column I. Barbara will therefore make a smaller loss by playing strategy I.

(c)   Allan: strategy I for $\frac{3}{5}$ of the time
strategy II never
strategy III for $\frac{2}{5}$ of the time
Barbara: strategy I for $\frac{2}{5}$ of the time
strategy II for $\frac{3}{5}$ of the time
strategy III never
Value of the game $= 4\frac{1}{5}$

5   North-west corner solution:

30		
70	30	
	20	60

Optimal solution $1 \to B; 30$
$2 \to A; 100$
$3 \to B; 20$
$3 \to C; 60$
Cost = 3180

6   (a)

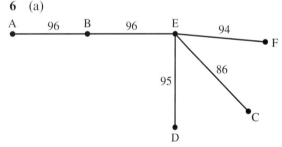

Arcs selected in the order AB, BE, EC, EF and ED.

(b)   (i)   934 km

(ii)   Use shortcut AF instead of ABEF.
Use shortcut FC instead of FEC.
Use shortcut CD instead of CED.
These shortcuts give a total saving of 213, giving an upper bound of 721 km.

(c) 635 km

(d)

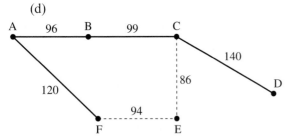

Using DE instead of CE increases the lower bound to 644. Route = ABCDEFA.

Examination style paper 2

1 x_{ij} = number of trucks sent from i to j

Z miles = distance covered

Minimise $Z = 50x_{AR} + 30x_{AS} + 70x_{AT}$
$$+20x_{BR} + 60x_{BS} + 10x_{BT}$$
$$+100x_{CR} + 40x_{CS} + 80x_{CT}$$

Since the problem is balanced, the constraints are:

$$x_{AR} + x_{AS} + x_{AT} = 35 \qquad \text{(A)}$$
$$x_{BR} + x_{BS} + x_{BT} = 60 \qquad \text{(B)}$$
$$x_{CR} + x_{CS} + x_{CT} = 25 \qquad \text{(C)}$$
$$x_{AR} + x_{BR} + x_{CR} = 30 \qquad \text{(R)}$$
$$x_{AS} + x_{BS} + x_{CS} = 45 \qquad \text{(S)}$$
$$x_{AT} + x_{BT} + x_{CT} = 45 \qquad \text{(T)}$$

$x_{ij} \geqslant 0$

2 (a)

	A	B	C	D	E	F
A	—	10	25	37	18	11
B	10	—	15	27	28	13
C	25	15	—	12	34	20
D	37	27	12	—	22	32
E	18	28	34	22	—	16
F	11	13	20	32	16	—

(b) Tour ABFEDCA

Total length (upper bound) = 98 km

(c) CA is not in the original network. This tour in the original network is ABFEDCBA.

(d) E.g. tour BAFEDCB

Length = 86 km

3 The maxim in route is SBET. The minimum length of an edge in this route is 13.

4 A gets 1–2 p.m. slot

B gets 3–4 p.m. slot

C gets 2–3 p.m. slot

D gets 4–5 p.m. slot.

Maximum audience exposure is 75.

5 (a) Optimal strategy for A:

choose strategy I for $\frac{7}{13}$ of the time

choose strategy II for $\frac{6}{13}$ of the time

Value of the game = $-\frac{3}{13}$

(b) Optimal strategy for B:

choose strategy I for zero time

choose strategy II for $\frac{19}{26}$ of the time

choose strategy III for $\frac{7}{26}$ of the time

6 (a) $1 \rightarrow A$, 25 truckloads

$1 \rightarrow B$, 20 truckloads

$2 \rightarrow C$, 18 truckloads

$3 \rightarrow B$, 15 truckloads

$3 \rightarrow C$, 2 truckloads

Cost = $550 \times £10 = £5500$

(b) Source 1 is fully used.

Source 2 does not need to be used at all.

Source 3 will only send 15 of the 17 loads it has available.

(c) $1 \rightarrow A$, 25 truckloads

$1 \rightarrow B$, 18 truckloads

$1 \rightarrow C$, 2 truckloads

$2 \rightarrow C$, 18 truckloads

$3 \rightarrow B$, 17 truckloads

Index